Bastioned Empyrean Bastion using Xeroxion Custoship

Vesna Nadya

Bastioned Empyrean Bastion using Xeroxion Custoship

First Edition October 2025

Written by Vesna Nadya

TABLE OF CONTENTS

CHAPTER 1

INTRODUCTION

1.1 INTRODUCTION TO CLOUD COMPUTING

In recent years, the computing industry has undergone a major transformation with the rise of cloud computing—a paradigm in which computing services are provided on-demand. Cloud computing offers several benefits, including elasticity, scalability, and cost efficiency. In this model, users can access computing resources based on their needs, without worrying about where the services are hosted, how the data is stored, or how the infrastructure is managed. Users pay only for the resources they consume, freeing them from maintenance, security, and other operational concerns.

This approach allows organizations to adopt new systems and features without significant upfront investment in software or infrastructure. Cloud services, managed by service providers, provide a flexible environment for developing applications and integrating new capabilities into existing systems. Compared with traditional computing models, cloud computing is more efficient in terms of cost, time, and resource management. Businesses can leverage software, platforms, and infrastructure from the cloud, paying only for what they use while avoiding the burden of administration.

As organizations migrate to the cloud, the security and privacy of information become paramount. Protecting sensitive data is essential since information is a valuable asset. While service providers are responsible for ensuring security, users benefit from a hassle-free experience, allowing them to focus on utilizing cloud resources effectively.

Cloud computing is an IT model for computing which includes all the hardware, software, networking and services to enable development and delivery of services through the internet. Cloud computing is becoming popular due to the undeterminable benefits offered to its users. In simple cloud computing is defined as storing and accessing data from remote storage using internet. The trend towards cloud computing started with grid computing which provided a virtual pool of computational resources but it is different from cloud computing. The National Institute of Standard and Technology (NIST) is an agency, which is meant for promoting innovations in technology defined cloud computing as a model for enabling ubiquitous, convenient, on-demand network access to a shared pool of configurable computing resources that can be rapidly provisioned and released with minimal management effort or service provider interaction. Kalloniatis *et al.,* [2014] has listed five main characteristics of cloud computing as:

- **On-demand self-service:** Cloud users can obtain additional resources including storage, servers and applications whenever needed without any human intervention and are self manageable.
- **Broad network access**: Cloud users can access the capabilities over network using standard mechanisms in different devices.
- **Resource pooling:** The resources are not dedicated to any specific client .It is based on multi-tenant model. The resources are pooled to serve multiple cloud users.
- **Rapid elasticity:** Capabilities can be elastically provisioned and released on demand.
- **Measured service:** Adopts pay per use method. Cloud users have to pay based on the usage of resources.

Cloud storage stores the user data on storage system maintained by a third party. This means that the user can store the data on remote location and can access it through internet. Data outsourcing to cloud storage is the recent trend and many organizations are moving into cloud because it allows the users to store their data on clouds and can access at anytime from anywhere.

This chapter outlines the fundamentals of cloud computing in section 1.2 and the service modes are explained in section 1.3. Section 1.4 discusses about the cloud deployment models. The basic architecture, data storage architecture and layered architecture of cloud computing are discussed in sections 1.5, 1.6 and 1.7 respectively.

The attractive features of cloud computing are listed out in section 1.7 ,section 1.8 presents the security attacks in the cloud and section 1.9 presents the popular Cloud Service

Providers (CSPs) available in the market. The final part of the chapter presents the motivation, objectives, research problem statement formation and the thesis organization from sections 1.10 to 1.13 respectively.

1.2 FUNDAMENTALS OF CLOUD COMPUTING

Cloud computing is a computing paradigm based on distributed computing, which is a sensational research platform. The distributed computing platform relies on numerous computing devices located in different locations, while the computing devices are interconnected for the sake of communication and data sharing. Hence, for a task to be accomplished, the scattered computing devices must coordinate among themselves.

The shared memory of the computing device is given access to all the processors for information sharing. Additionally, the computing device is equipped with the private memory. One of the most popular computing models is the grid computing, which forms the base for the evolution of cloud computing. The main objective of grid computing is to share the workload of the computing devices. Grid computing supports in utilizing the computing resources through the concept of virtualization, as cited by Fehling *et al.,* [2014], Furht [2010].

Cloud computing is the evolutionary model of grid computing, which utilizes the internet and the data management is carried out by several remote servers. The cloud computing environment provides three vital services such as Infrastructure as a Service (IaaS), Platform as a Service (PaaS) and Software as a Service (SaaS). The cloud user does not need any special requirement to utilize the cloud services, but the only requirement is the network connection. As mentioned by Fehling *et al.,* [2014] and Durao *et al.,* [2014], the cloud user is charged for the services being availed.

Today's world deals with enormous amount of data each and every second of time. Most of the data are claimed to be confidential, as they comprise personal, health and financial information. The amount of data escalates with respect to time and it is very difficult for the organizations to store and manage the data. Hence this idea is welcomed by numerous mid-scale and small-scale industries, as the issue of licensed version purchase, data management are handled effectively for a nominal payment. Additionally, the performance of computing is better, due to the presence of numerous remote servers, processors, memory and bandwidth. The service modes of cloud computing are presented in the following section.

1.3 SERVICE MODES OF CLOUD COMPUTING

The cloud computing environment provides a wide range of services to the cloud users and the most popular broad modes of services are IaaS, PaaS and SaaS as explained below. The service modes of cloud computing are depicted in Figure 1.1.

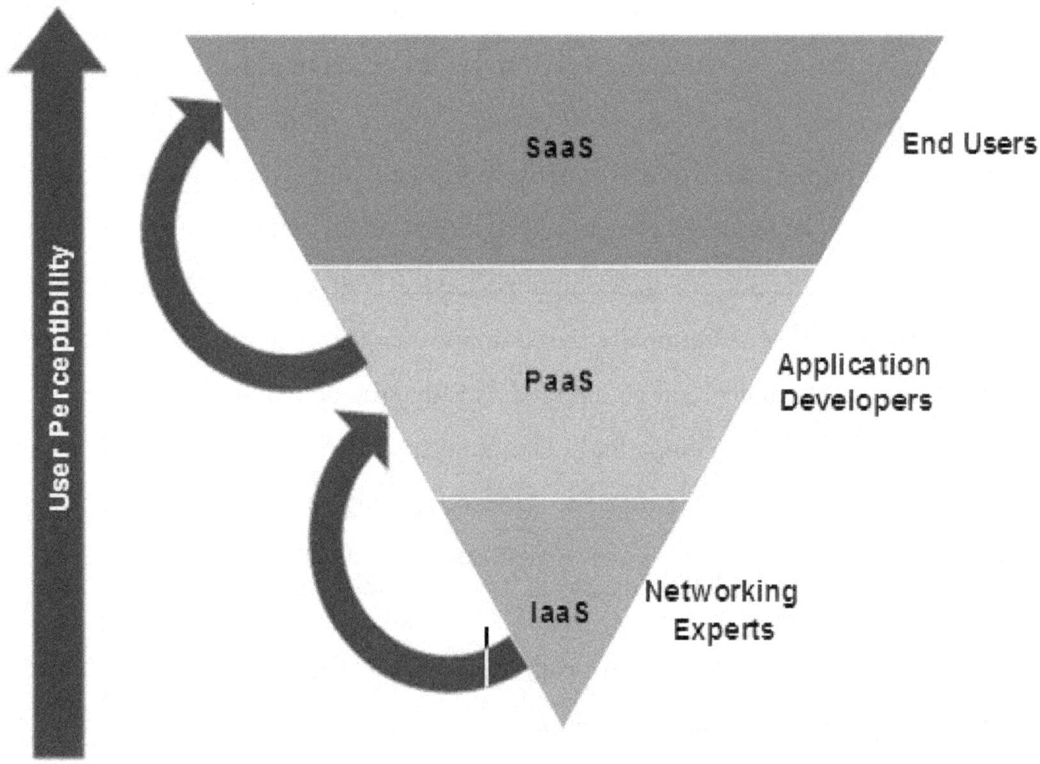

Figure 1.1: Service Modes of Cloud Computing

1.3.1 IaaS

Iaas as a model offers the infrastructure to run applications with basic storage and computing capabilities as standardized services over the network. IaaS is the foundation to all cloud services. IaaS provides the central storage and computing terminals through the network to the cloud users. This service is made possible by deploying numerous servers, storage resources and other necessary systems in the cloud computing environment. Though the features of cloud computing seems to be similar to standard enterprise computing, the cloud computing differs from it by providing the entire infrastructure as a service.

The users are allowed to customize the service based on the user requirement as well. Thanks to the concept of virtualization that makes the abstraction, encapsulation and better service provision of hardware resources possible. The main advantages of this service are its flexible payment model and the Quality of Service (QoS). The payment is fixed and follows the pay-as - you-go pricing model. As discussed by [5] some of the IaaS providing CSPs are Amazon EC2, GoGrid, iCloud and Microsoft Azure DC.

1.3.2 Platform as a Service (PaaS)

PaaS provides the software platform to the cloud users upon which the software based services can be built in such a way that it provides a programming platform for developers. The PaaS is provided to the cloud user with the major components such as integrated Operating System (OS), middleware, application software and so on.

All these services can be accessed by the cloud users and this service is mostly utilized by the software developers, as PaaS provides most of the services for building applications. This service enables the cloud users to utilize several automated tools, which helps in performance enhancement and reducing the time consumption. As presented by Kavis [2014] some of the sample PaaS supporting CSPs are Google App Engine, force.com.

1.3.3 Software as a Service (SaaS)

SaaS offers the cloud users with the software application built in advance and is hosted as a single application that can support a wide range of users. In SaaS model, the users can use the software for rent instead of purchasing it. The applications are accessible from various client devices through web browsers. Basically, the components being involved in the applications can be utilized or removed based on the user requirements. Hence, the concept of customization is completely supported and the service is provided under the policy called multi-tenancy.

The multi-tenancy concept helps multiple users to access a single application by creating instances. Hence, the user can enjoy the feel of private utilization of service. Besides, the cloud users are freed from software updations and management, the customers do not manage or control servers, networks, operating systems, storage and even underlying infrastructure. Some of the sample SaaS offering CSPs are Salesforce.com, Microsoft office 365, workday, as discussed by Cusumano [2010] and Wu *et al.,* [2011]. Hence, the most fundamental ideas about cloud service models are discussed in this section and the following section gives an overview of cloud deployment models.

1.4 CLOUD DEPLOYMENT MODELS

The cloud models are deployed in four significant categories and they are public, private, hybrid and community clouds. All these deployment models are summarized in the following section

1.4.1 Public Cloud

Public clouds generally owned by organizations, government or some combination of them offer services that can be rendered by all users . In case of public cloud, the third party Cloud Service Provider (CSP) provides the cloud services to the cloud users. Mostly this kind of cloud allows the public users to utilize the services through network. The main advantage of the public cloud is the scalability, while security is the major concern of public clouds.

Figure 1.2: Representation of Public Cloud

The main reason for this is that the cloud service can be accessed by the public without any security baseline. The user and service provider works on the basis on the Service Level Agreement(SLA), which is a contract being signed between two parties. The SLA documents the services provided by the provider along with the measurable aspects for the services. The public cloud concept is depicted in Figure 1.2.

The public cloud offers services to the general public or a large industry group and is owned by an organization selling cloud services. The designated service provider will provide public clouds and all customers will share the same infrastructure with limited configurations and security controls. In this case, the computing is based on massive scale offerings to the general public. The infrastructure is sited on the grounds of cloud service provider, which is maintained by the cloud infrastructure. However, the public cloud users are not trustworthy as the services are open to public and the users are not bound to any service agreements. The major advantage of public cloud is that providers only charge for the services used by the users. As in Ren *et al.,* [2012], Fehling *et al.,* [2014] some of the famous cloud services provided by public clouds are Amazon EC2, Amazon S3, salesforce, Google App Engine and Force.com.

1.4.2 Private Cloud

Private cloud infrastructure is employed absolutely for an organisation. Private clouds are meant for a specific group of people, which could be an enterprise. It may be managed by an organization or a third party and may exist on premise or off premise. As the services are accessible only with a particular group of people being associated with the enterprise, security is enhanced and also the QoS is improved in the private clouds. The cloud infrastructure lies under the control of an enterprise, such that the data control is made easier. The concept of private clouds is depicted in Figure 1.3.

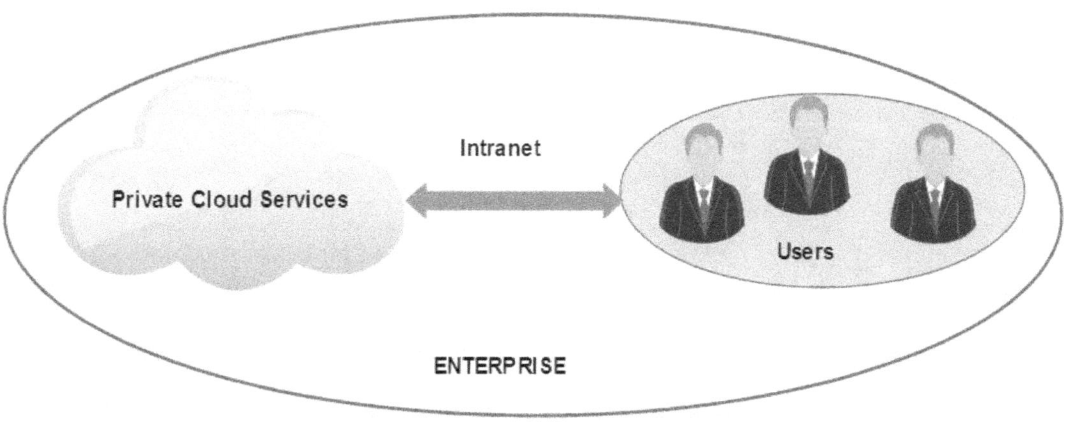

Figure 1.3: Representation of Private Cloud

Usually, the private clouds are set up in the enterprise datacenter. However, the scalability of private clouds is based on the potential of the datacenter. Usually, the private cloud infrastructure is managed by a single enterprise and the main objective for the enterprise to employ private cloud is to improve the exploitation of the available resources and to reduce the security related issues. As mentioned by Dillon *et al.,* [2010*a*], Doelitzscher *et al.,* [2011], mostly private clouds are employed by universities, educational institutions and Information Technology (IT) enterprises.

1.4.3 Hybrid Cloud

Hybrid clouds are the clouds which club the functionalities of both the public and private cloud models, such that the advantages of both the public and private models are inherited by the hybrid cloud that remain unique entities but are bound together by standardized or proprietary technology that enables data and application portability. The main purpose of a hybrid cloud is to provide flexibility, scalability and additional resources whenever needed without sacrificing security and performance of local infrastructure. The hybrid cloud model is shown in Figure 1.4.

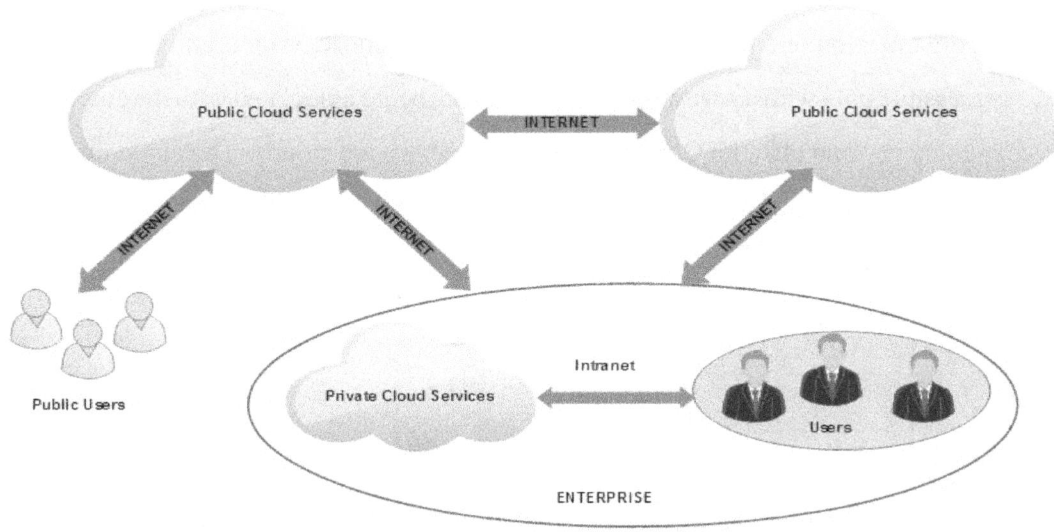

Figure 1.4: Representation of Hybrid Cloud

Usually, the hybrid clouds employ more of private clouds and when the resources are limited public clouds are accessed. Hence, the resources from both the private and public clouds are utilized. All the conditions about private and public cloud utilization are mentioned in the SLA. Mostly, the enterprises employ hybrid model for enhancing the resource provisioning and the QoS.

1.4.4 Community Cloud

Community cloud works by creating a community and provides specialized services based on the community. It shares the infrastructure among a specific community and is managed internally by organizations or by third party and may exist on premise or off premise. The members of a specific community can access a specific type of service, share data and communicate among themselves. Each and every community has its own cloud data and services. The main issue in community cloud is as the data is located at one place strict security controls should be used inorder to store the data, otherwise it might be accessible to others. The community cloud is depicted in Figure 1.5.

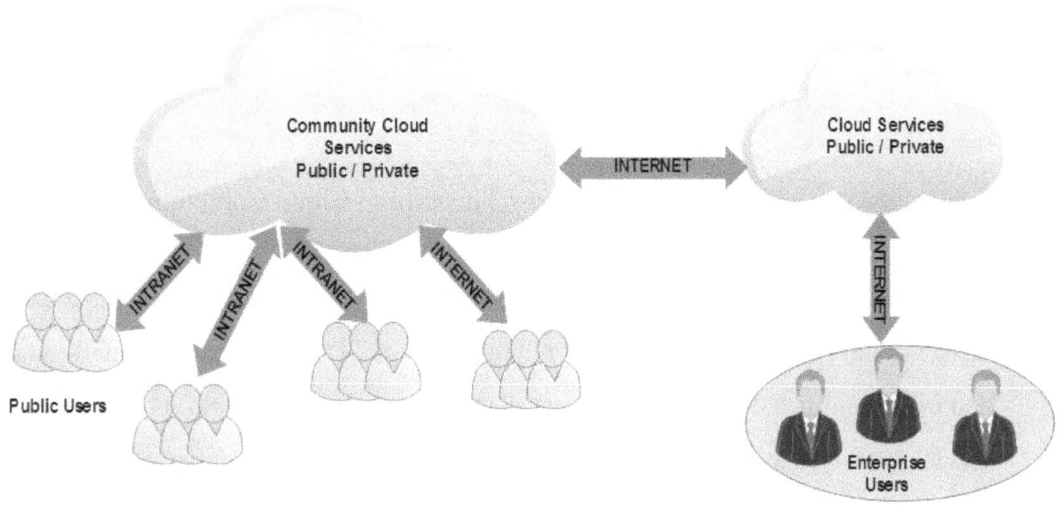

Figure 1.5: Representation of Community Cloud

Usually, multiple organizations combine together such that a single cloud infrastructure can be shared on the basis of certain policies and constraints. This kind of cloud infrastructure can be provided by a third-party vendor or any of the organizations in the community, as mentioned by Marinos and Briscoe [2009].

The basic blocks of cloud computing model are sketched with details about public, private, community and hybrid cloud models. The following section outlines the basic architecture of cloud computing.

1.5 BASIC ARCHITECTURE OF CLOUD COMPUTING

The basic architecture of cloud computing is based on three important entities and they are cloud service users, CSPs and the Internet for communication. The cloud service can be accessed with the help of any computing device such as Personal Digital Assistant (PDA), mobile phones, tablets and so on. The service is provided by to the cloud service users by the CSP via the internet and is represented in Figure 1.6.

Figure 1.6: Basic Architecture of Cloud Computing

The cloud computing architecture can be observed with three important perspectives such as front end, back end platform and a network host. The front end is usually visible to the end client and it is the opening point of an application or service. For instance, the cloud services can be accessed with the help of any user terminal as shown in Figure 1.6.

On the contrary, the back end of the cloud architecture is invisible to the end users, however it is potential. Usually, the back end is formed by servers, data storage, processing systems and so on. A central governing authority manages the entire system, which includes communication between entities, data transmission and storage.

All these processes are taken care of by the middleware, which is meant for bringing in interoperability by employing standard protocols. Besides this, the computational power of the server is completely utilized by means of a unique concept called Virtualization. The concept of Virtualization is based on two significant entities called Virtual Machine (VM) and VM Monitor (VMM).

The VMM software supports the execution of multiple Operating System (OS) on a single hardware at a specific time. The VM is a software through which an OS can be executed. One of the best examples is the Google Apps through which the required application is accessed with the help of computing device through the internet.

The basic entities involved in the cloud computing architecture are outlined in this section and the following section explains the data storage architecture of cloud computing.

1.6 DATA STORAGE ARCHITECTURE OF CLOUD COMPUTING

Data is the important asset in business . Some businesses handle data for millions of customers , and need the security and integrity of data for a long period of time. Data created by individuals or business must be stored so that it is accessible for further processing.

Inorder to reduce the burden of data management ,businesses are moving the data to cloud. Cloud primarily refers to storing data to a storage system maintained by a third party. The NIST formulates an exclusive architectural framework for cloud computing, which considers the cloud users, actions and the operations performed in cloud computing.

As cited by Liu *et al.,* [2011], Bohn *et al.,* [2011] the architectural framework clearly depicts the high-level architecture, which helps in understanding the requirements, functionalities and standards of cloud computing. The architectural diagram shown by NIST is presented in Figure 1.7 and all the entities involved are explained in the subsections.

1.6.1 Cloud Service User

The cloud service user is the main entity meant for utilizing the cloud service. A cloud service user can be an individual or an enterprise meant for consuming the cloud service from the CSP. Cloud service user relies on the cloud for data maintenance and computation. In order to utilize a cloud service, the cloud service user navigates through the service directory being advertised by the CSP for selecting the suitable service.

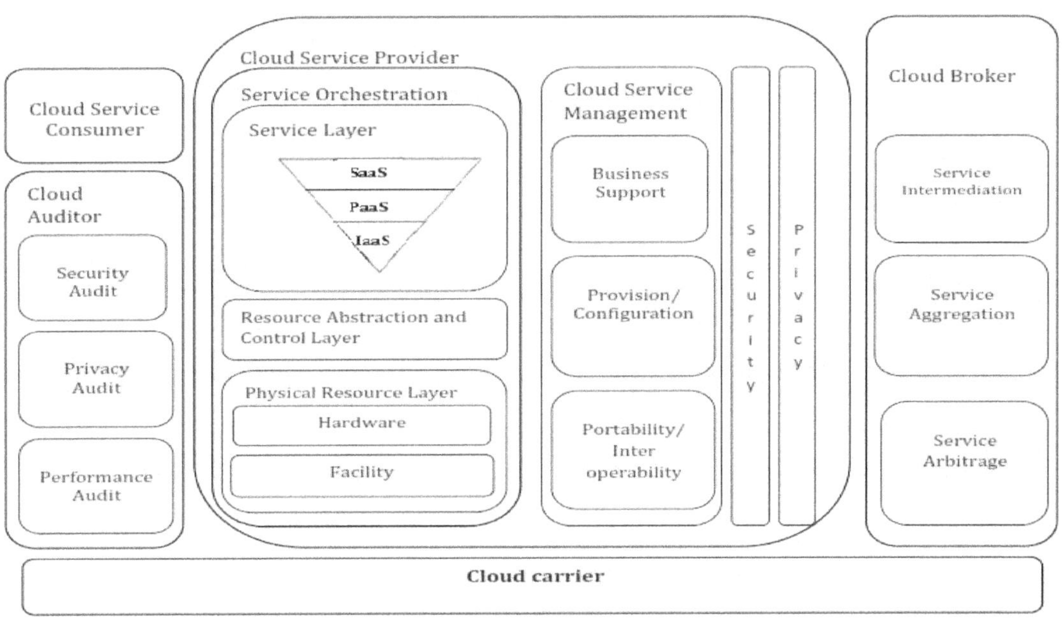

Figure 1.7: NIST Cloud Architecture

Once the suitable service is detected, the cloud service user signs the SLA with the CSP for service utilization. Once the service is accessed from the cloud server, the cloud service user is subjected to a nominal fee, as per the SLA.

1.6.2 CSP

A CSP could be an enterprise again meant for providing service to the cloud service users. The CSP is responsible for ensuring the service availability at all times to the cloud service users. Certain other important functionalities of CSP are it maintains the cloud components meant for service provisioning and executes the software to provide access to the cloud services for the users. The cloud services are hosted in a data center that can be accessed by cloud service users through internet.

1.6.3 Cloud Auditor

Cloud auditor is an important entity that inspects the cloud services and governs all the activities being performed by both the cloud service users and the CSP. This operation is carried out by performing audits and the compliance of standards is checked. Cloud audit is meant for analysing the security level, privacy, performance, QoS and so on.

1.6.4 Cloud Broker

As the concept of cloud computing is vast, the cloud service users may find it difficult to find the required service from a bulk of services provided by the CSP. In this case, a cloud broker comes into picture by suggesting the appropriate service to the cloud service user from a list of services.

The cloud service user requests the cloud broker to suggest the most appropriate service, rather than navigating through the list of services supported by the CSP. This idea enhances the choice of service and the cloud broker takes the responsibility of selecting and providing services to the cloud service users, while acting as an intermediate entity between the cloud service user and CSP.

1.6.5 Cloud Carrier

Cloud carrier is another important entity that is responsible for ensuring the connectivity between the cloud service users and CSP. Cloud carrier delivers the services to the cloud service users by utilizing the networking technology. This makes it possible for the cloud service users to access the cloud services from any of the computing devices such as PDAs, mobile phones, tablets or computer.

Hence, this section discusses about the major components involved in the cloud computing architectural framework and the following section gives an idea of the layered architecture of cloud computing.

1.6.6 Layered Architecture of Cloud Computing

The layered architecture of cloud computing describes the NIST cloud architecture in a simplified way. The cloud architecture is segregated into three different layers and is diagrammatically presented in Figure 1.8.

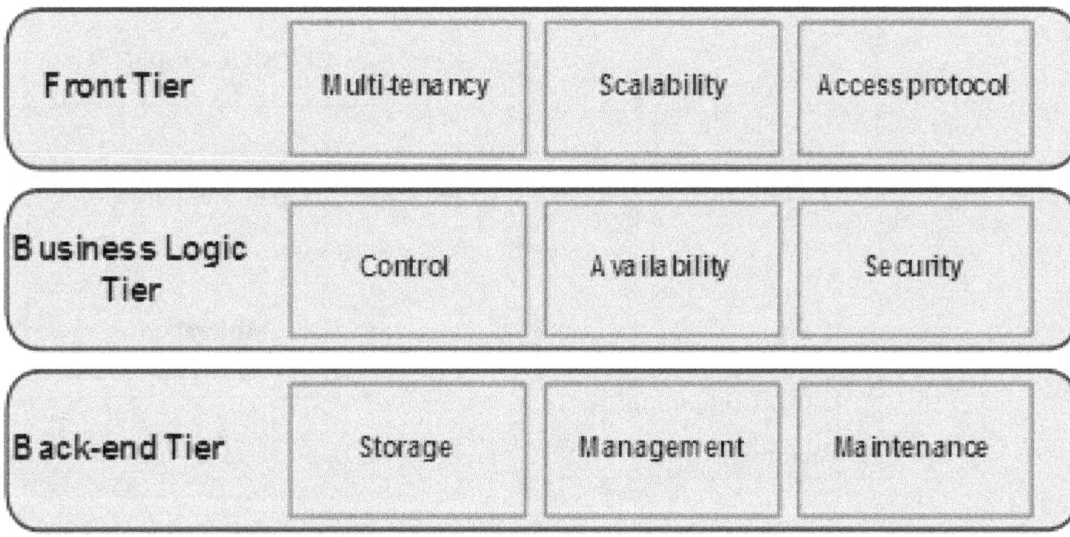

Figure 1.8: Layered Architecture of Cloud Computing

The front end is meant for enabling the communication between the cloud service user and the cloud server. The user terminal can be different as mentioned in the previous section. This layer focuses on multi-tenancy by handling separate instance for every user. Additionally, scalability is supported by employing different techniques.

The business logic layer is the main tricky layer which supports several functionalities and all the higher level procedures are incorporated in this layer. This layer concerns to provide security while ensuring availability and reliability. This layer has a complete control over the cloud data.

The back-end tier deals with the physical storage equipments and considers different protocols to deal with the physical data storage. This tier involves managing and maintaining the stored data, while increasing the storage efficiency.

Hence, the tiered architecture of cloud computing is discussed and the forthcoming section summarizes the attractive features supported by the cloud computing environment.

1.7 ATTRACTIVE FEATURES OF CLOUD COMPUTING

The cloud computing paradigm has shown numerous advantages when compared to the traditional computing environments. Some of the noteworthy features of cloud computing are depicted in Figure 1.9.

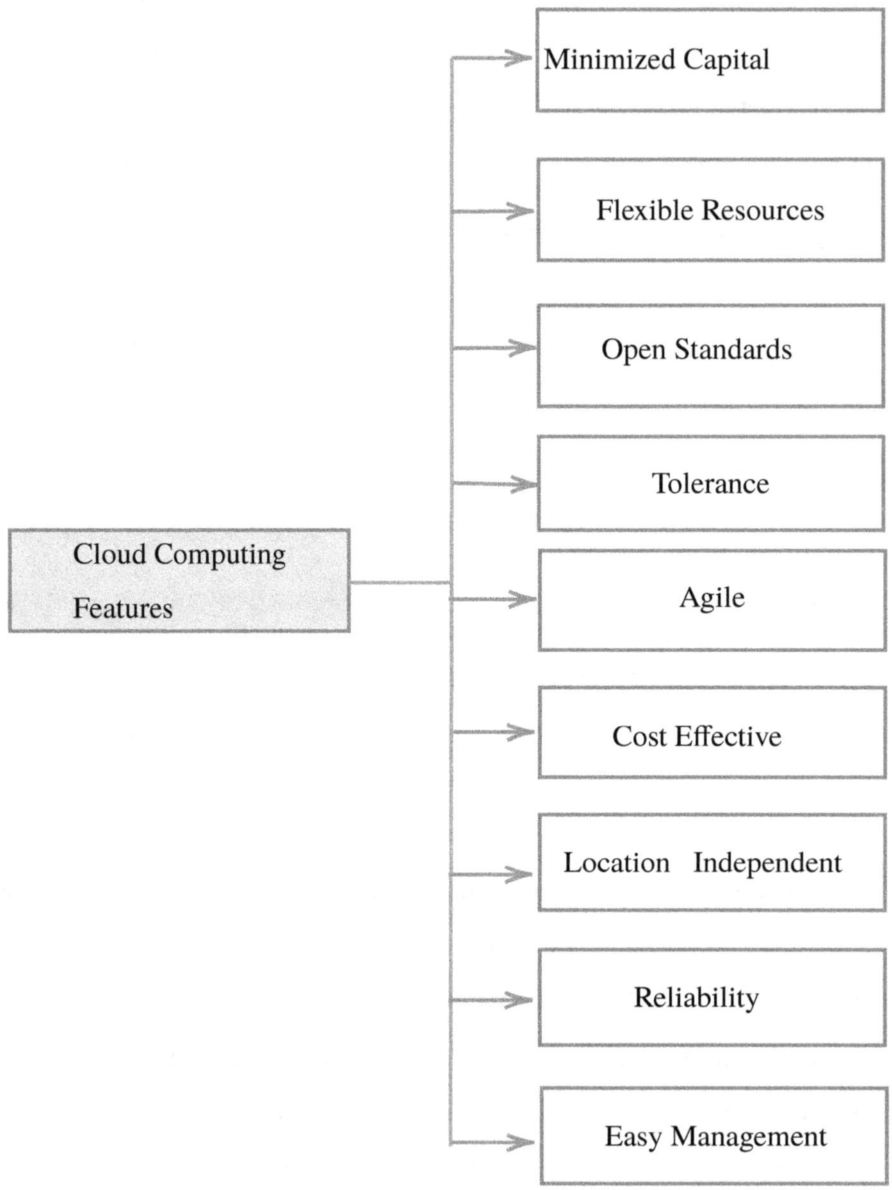

Figure 1.9: Features Of Cloud Computing

1.7.1 Minimized Capital

As the cloud computing provides all the necessary resources right from infrastructure to storage, there is no need for investing more on purchasing licensed software or storage equipment upgradation. Among all kinds of cloud, public clouds are even cheaper as all the resources can be accessed online. Though security is the main concern here, public clouds can be utilized for insensitive data.

The cloud service users are charged for the services being consumed and the users are prompted to pay only for the utilized resources. The cloud service users are charged on the basis of the signed agreement. This lowers the capital and encourages the beginners. The users can come out of the service whenever needed, as per the agreement.

1.7.2 Flexible resources

In most of the cases, a software application is undergone budget analysis, which roughly estimates the required storage space, processing requirements and other necessitates. This is to get rid of the last minute emergency and this scenario is completely avoided by the cloud computing. This is because the resources can be scaled up or down, whenever required. Hence, the progress in the software development does not get affected because of the resource constraints. This idea helps in handling the deadlines effectively and avoids the delay of work because of resource constraints, even though the planning is not perfect.

1.7.3 Open Standards

Mostly, all the abilities of cloud computing work is on the basis of open standards. The open source software is termed as the computer software, which is monitored by a software license in the public domain and it permits the users to exploit, modify and enhance the software. Open source software is the base of the cloud solution and is vital for its better growth.

1.7.4 Tolerance

The main objective of CSPs is to arrive at a sustainable architecture for ensuring stability. In order to make it happen, the CSPs spend a substantial amount. On the other hand, it is highly challenging for the enterprises to maintain a stable architecture, as a huge investment has to be made. Hence, the enterprises suffer from different failures such as single point of failure and inadaptability.

The enterprise should be able to cope up with the changing and advanced trend set of technology requirements. However, the cloud computing architecture supports the enterprises by controlling the issue of single point of failure and provides better adaptability.

1.7.5 Agility

Agility improves user's ability to re-provision technological infrastructure resources. With cloud computing, agility is brought in to the enterprise, such that the resources in technological infrastructure can be upgraded easily with no hassles. As cloud computing is agile, it can adapt to the changes effectively.

1.7.6 Cost Effectiveness

The cloud computing is cost effective by all means. The cloud infrastructure helps in minimizing the cost to be incurred by the enterprise as capital. Besides this, the cloud users utilize the cloud services, which are technologically advanced and upgraded. Hence, the users are relieved from the concern of constant updation and management of the utilized services.

1.7.7 Location Independence

The cloud computing provides a way for the cloud users to utilize the cloud services from any location and at any time. Hence, the cloud user can access any cloud service from anywhere at anytime through any computing device. The one and only requirement is the network connectivity.

1.7.8 Reliability

The cloud services are considered to be reliable, as the availability of services is not affected because of the distributed data storage pattern. Additionally, the cloud services are replicated in order to ensure data availability as well. This avoids data loss and the availability in turn provides the reliability.

1.7.9 Easy management

As the complete management of cloud services is taken care of by the CSP, the cloud users or the enterprise is released from the headache of managing the services. All the services are maintained in the cloud server and not in the local machine, which minimizes the complexity of management.

Usually, the cloud computing environment is utilized by individual cloud service users, start up businesses, mid and small scale industries, enterprise businesses and so on. Apart from these benefits, the cloud computing environment provides certain other benefits such as limitless storage, backup, software integration, accessibility, faster deployment and so on.

- **Limitless Storage:** As the cloud computing model is flexible and scalable, the users are not limited with respect to service utilization. The limitless storage is made possible by the concept of virtualization. This eliminates the need of concerning about the available storage space and storage upgradation.

- **Data Backup and Recovery:** As all the cloud data are stored in the remote cloud server, performing back up is quite easy. Additionally, remote data storage helps in easy data recovery too. For instance, in case of any natural calamities the storage is prone to single point of failure. In such scenarios, the remote cloud storage preserves data with data duplication as well. Hence, the data recovery can be done easily.

- **Software Integration:** The software integration and all cases of software upgradations are done automatically by the CSP. Hence, the cloud service users can enjoy the recent technologies by making reasonable payment.

- **Accessibility:** Once, the cloud users sign the contract with the CSP, the cloud users can access the cloud services right from any part of the world, provided the user has got internet connection.

- **Faster Deployment:** The most attractive benefit of cloud computing is that an enterprise can be made functional in a matter of few hours. The only demand is that the complete system must fall under the cloud computing environment.

Hence, this section presents all the benefits provided by the CSP to the cloud users and the challenges faced by cloud computing are listed below.

1.8 CHALLENGES IN CLOUD COMPUTING

Zhang *et al.,* [2010] stated that though cloud computing environment ensures time and cost efficiency, there are several challenges placed in front of cloud computing and this section gives an overview of the most serious challenges, as depicted in Figure 1.10.

1.8.1 Privacy

Privacy refers to the ability of preserving confidentiality on stored information and preventing it from misuse. Privacy is the most serious challenge faced by cloud computing environment as the data is with the third party(CSP). The data stored in the cloud server is completely taken care of by the CSP, which indirectly means that the CSP has the complete control over the data. Hence, the CSP may intrude through the data for gaining information and hence, privacy is the main concern of any data owner. Irrespective of the presence of SLAs, privacy breach is still a critical issue to be addressed. Auditing mechanisms can be used for monitoring CSP for data misuse inorder to preserve privacy. The main challenges of privacy for cloud computing are

- Complexity of risk assessment in a cloud environment
- Emergence of new business models and their implications
- Achieving regulatory compliance

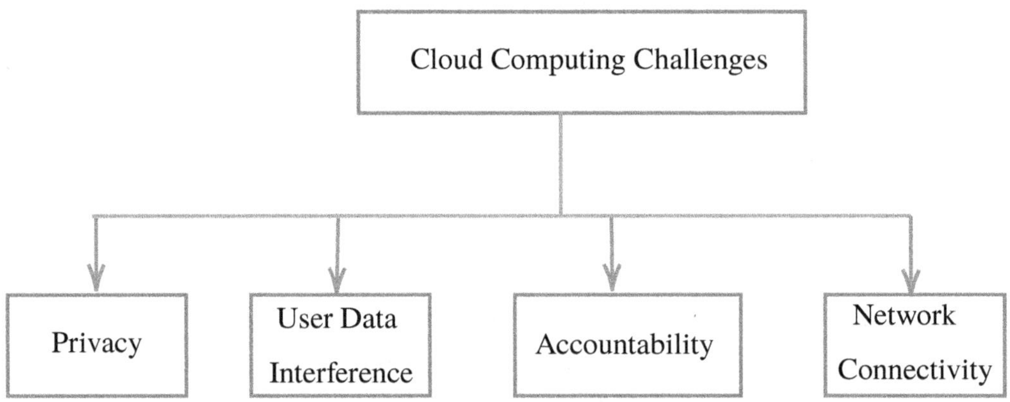

Figure 1.10: Cloud Computing Challenges

1.8.2 User Data Interference

User interference is another serious threat to the cloud computing environment. As mentioned already, cloud computing offers limitless storage space for data storage and the data duplicates are created in order to increase the data availability. However, when the data is deleted, the changes may not reflect on the duplicates. These data can be utilized for gaining information and the data is sold in some cases. Hence, user data interference may occur due to the enormous data storage.

1.8.3 Accountability

It is always a million dollar question about the normal data access during the failure of the cloud server. Additionally, no sole responsibility is allotted for the data loss and the data misuse. The accountability is ensured when the CSP protects data and clarifies about whom and all can process the data without considering the location.

1.8.4 Network Connectivity

As the life of cloud computing depends on the network connectivity, and in case of network failure, the services provided by cloud are completely inaccessible. Hence, the cloud services are not suitable for the places with poor network connectivity.

Hence, this section discusses about the four major challenges faced by cloud computing and the following section presents the most popular cloud service providers in the market.

1.9 SECURITY VULNERABILITIES AND THREATS IN THE CLOUD

The way data is stored, used and shared is transformed with the introduction of cloud computing. Many organizations and individuals have already improved the efficiency of their operations while reducing the IT costs. With the ease of managing the data, a host of new security and challenges has also been introduced.

The key security vulnerabilities of cloud computing are: data threats, cloud API vulnerabilities, malicious insiders, insufficient identity, credential and access management, shared technology vulnerabilities, provider lock-in, weak cryptography and vulnerable cloud services and are depicted in figure1.11

1.9.1 Data Threats

Data stored in the cloud is susceptible to loss, breach or any other damage as a result of human errors, poor security practices, vulnerabilities in applications etc. It is the responsibility of the CSPs to ensure the safety and security of the data in the cloud. The confidentiality, integrity and availability of the data should be ensured so that the data owners and data users are both benefitted.

1.9.2 Cloud API Vulnerabilities

Cloud developers should implement strong controls over APIs. APIs are used to allow the users to interact with the cloud based services. The vulnerabilities in APIs could have an impact on the security of cloud orchestration, provisioning, management and monitoring. So APIs should be designed with utmost care so that no fraudulent attempts can be made through it.

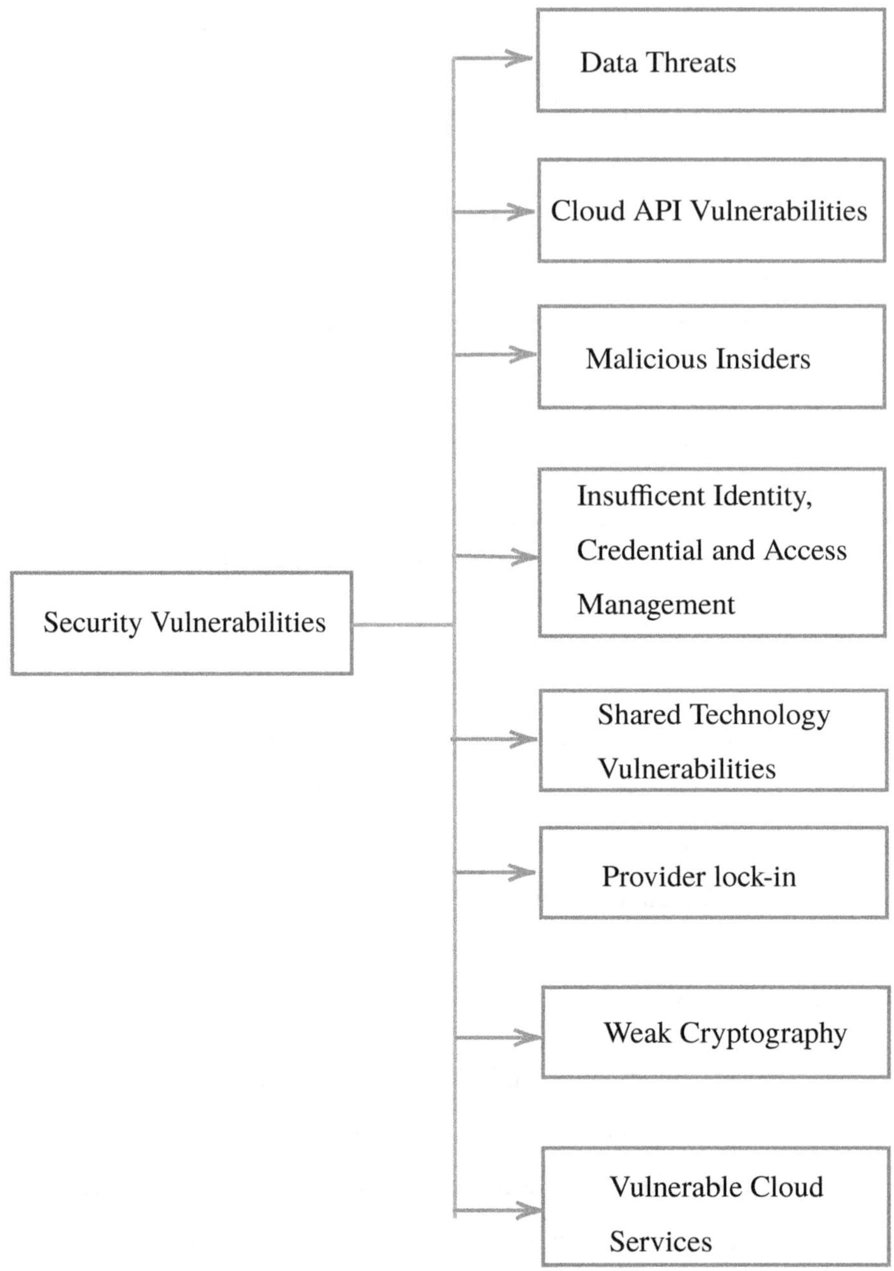

Figure 1.11: Security Vulnerabilities

1.9.3 Malicious Insiders

Legitimate users sometimes behave wrong. They have the full control of the data. They can make any changes to the data, leak them out or make use of that data in an illegal manner.

1.9.4 Insufficient Identity, Credential and Access Management

Attackers or bad actors masquerading as genuine users or operators could make use of the data as they wish. They could even release software that appears as such as the original ones.

So the insufficiency in dealing with identity, credential and access management can facilitate the unauthorized access to data resulting in catastrophic damage in the organizations or the end users.

1.9.5 Shared Technology Vulnerabilities

Cloud uses shared technologies like virtualization., cloud orchestration etc. Any vulnerability in these technologies makes way for the hackers to gain access to the data. For instance, a weakness in the hypervisor can allow the hackers to gain control over the virtual machines or even the host.

1.9.6 Provider Lock-in

It's quite common for cloud service providers to make their clients highly dependent on them with higher switching costs. Due to the cost incurred in switching from one provider to another, users tend to get locked in particular providers even if they are not able to provide all the services demanded by the users.

1.9.7 Weak Cryptography

To secure the data, cryptographic algorithms are used. So before taking the data to cloud, it must be secured strongly, these algorithms need to be made strong. The hackers use efficient decoding techniques to hack the information.

1.9.8 Vulnerable Cloud Services

The cloud services are distributed and so they have little protection against each other. The vulnerability in one cloud service can be used to gain unauthorized access to the data of legitimate users. Strong architecture are recommended for isolating user's operations in the cloud.

1.10 SECURITY ATTACKS IN THE CLOUD

Businesses today are migrating from the traditional way of computing to cloud computing as they provide numerous benefits to the organizations. Amidst all the attractive features cloud computing faces many challenges among which security gets the prime importance. Some of the common security issues faced by the cloud adopters are: network availability, privacy of the data, control over the data, cloud provider viability, disaster recovery and business continuity, systems vulnerabilities and risk of common attacks, regulatory or legislative compliance.

The cyber attacks against cloud computing are done with the goal of getting access to the user data and preventing the genuine users from accessing the services. The most common types of cyber attacks are: cloud malware injection attacks, abuse of cloud services, denial of service attacks, side channel attacks, wrapping attacks, man-in-the cloud attacks, insider attacks, account or service hijacking, advanced persistent threats spectre and meltdown.

1.10.1 Cloud Malware Injection Attacks

To take control of the information of an user, these attacks are performed. This is done by adding an infected service implementation module to SaaS or PaaS solution. Once the system is deceived, all the incoming requests will be redirected to the hacker's module thereby initiating the execution of the malicious code. After this, the attacker can do any malicious activity such as manipulating, stealing the information. Cross-site scripting, SQL injections are the most common forms of malware injection attacks. In cross site scripting, malicious scripts are added to a vulnerable web page by the hackers.

1.10.2 Abuse of Cloud Services

Cheap cloud services are used to arrange Denial Of Service (DOS) attacks and brute force attacks on target users.

1.10.3 DOS attacks

The attacks are designed to overload the system and deny the services to the genuine users. When zombie machines are used in large numbers to attack, then the attack becomes Distributed Denial of Service attack.

1.10.4 Side Channel Attacks

These attacks are arranged by placing a malicious virtual machine on the same host as the target virtual machine. Usually the host's implementations of cryptographic algorithms are targeted. A secure system design is a good solution for side channel attacks.

1.10.5 Wrapping Attacks

Wrapping attack is an example of man-in-the middle attack. As the cloud users connect to the services through a web browser. Cloud computing is vulnerable to these attacks. By this, user credentials can be accessed and used maliciously.

1.10.6 Insider Attacks

Legitimate users can turn against the organization sometimes. These users purposefully violate the security policy. By providing different levels of access to cloud services, these attacks can be minimized.

1.10.7 Account of Service Hijacking

Account hijacking is done when the users credentials are accessed. There are various techniques to get this done including spyware, cookie poisoning etc. Once the credentials are gained access , then the service or account can be used as the hacker's wish to.

1.10.8 Advanced Persistent Threats

Without the knowledge of the legitimate users the hacker's steal the data stored in the cloud or exploit the services. Hacker's can work efficiently with the data center networks and use the network traffic for any malicious activity.

1.10.9 Spectre and Meltdown

These are new kinds of attacks which affects all devices ranging from smart phones to servers. With the help of malicious javascript code, data could be read from memory. These attacks pave way for side channel attacks as the isolation between applications is broken down. By this an attacker is able to access a system through unprivileged login and read information from the kernel.

1.11 POPULAR CLOUD SERVICE PROVIDERS

The CSPs invest so much of money to build a high quality cloud infrastructure. Hence, the technological giants such as Amazon, Microsoft, Google, IBM provides several flavours of cloud services to the users. The users can opt for the required services, which shall be provided by the CSP in turn. The most popular CSPs are depicted in Figure 1.12.

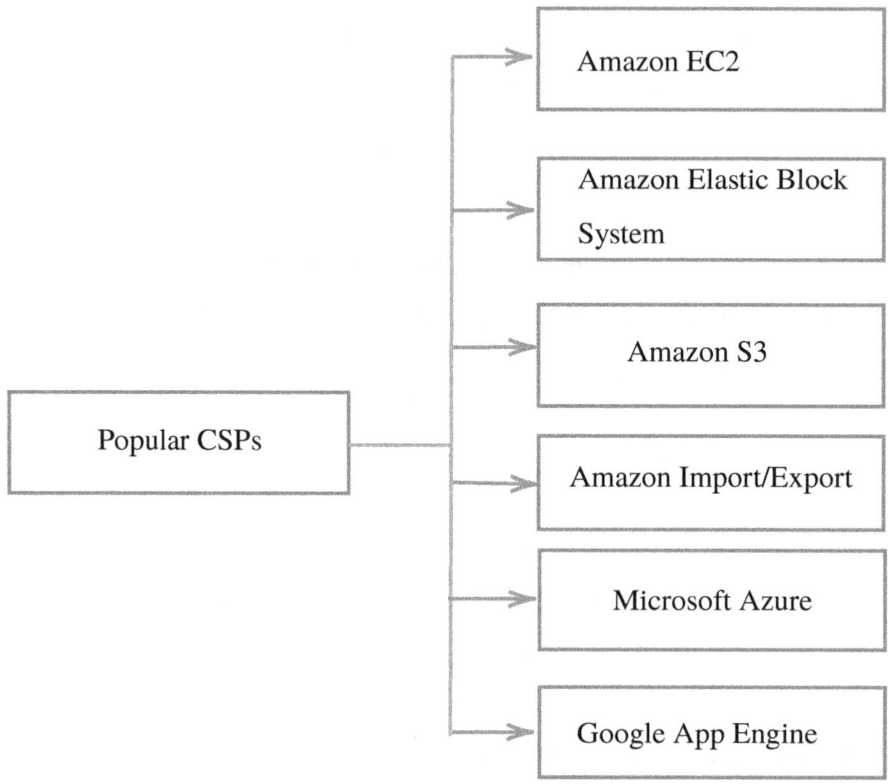

Figure 1.12: Popular CSPs

1.11.1 Amazon EC2

Amazon Web Services (AWS) is one of the most famous CSPs, which provides a numerous services to the cloud users on demand. The cloud service users make use of the cloud services through Hyper Text Transfer Protocol (HTTP), which is an application layer protocol. The web service access is rendered to the users by employing the web service access protocols such as Representational State Transfer (REST) and Simple Object Access Protocol (SOAP).

The Amazon Elastic Cloud Compute(EC2) cloud permit the cloud users to exploit the cloud server instances by means of the API. The instances formed by EC2 function like the virtual machines, which are executed on Xen virtualization engine. When the instance is instantiated and initialized, the cloud users are allowed to upload and manage the software.

In the same way, the data may be replicated and uploaded to improve the availability. These instances are located in several EC2 areas and the performance of the EC2 cloud is governed by the Amazon CloudWatch, which can assess the utilization level of Central Processing Unit (CPU), upload and download bytes and so on.

1.11.2 Amazon Elastic Block System (EBS)

Amazon Elastic Block System popularly known as Amazon EBS, permits to append the storage volumes ranging from 1GB to 1TB into EC2 instances. The significant feature here is that the EBS has a snapshot function, which can save the present status of the EBS and allows data restoration or replication.

If granted access, such snapshots can also be shared with other users, allowing other users to access and the working environment is made simpler. New users are offered with 30GB storage space, 2 million I/O and 1 GB of snapshot storage for every month up to a year.

1.11.3 Amazon Simple Storage Service (Amazon S3)

Amazon Simple Storage Service also called as S3, is a storage space provided by Amazon to the cloud users and this storage is termed as buckets. The S3 supports to store numerous data objects up to 5 TB and it is usually combined with Amazon EC2 or some other services. The durability and reliability are provided by S3 along with a special Amazon S3 Encryption Client library, which supports to encrypt data.

In addition to this, Reduced Redundancy Storage (RRS) is also provided for minimizing the costs by duplicating data in certain important locations. Though this is meant for less significant data, it is trickier than the S3. The new users of S3 are gifted with 5 GB of Amazon S3 storage, 20,000 Get Requests, 2,000 Put Requests, and 15GB of data transfer for every month till one year.

1.11.4 Amazon Import/Export

Amazon Import/Export is a service provided by Amazon that forwards the storage device to the address of the user. This is to ensure that the large amounts of data can be transferred directly into Amazon S3 or EBS. This kind of direct transfer is attained quicker than the transfer with internet. However, the shipping cost is more.

1.11.5 Microsoft Azure

Microsoft Azure provides an environment resembling windows platform, that can provide the cloud users to run the applications and store the data on the cloud server. The data services and infrastructure services are provided to the cloud users by means of Structured Query Language (SQL) and .NET framework. The data are saved in relational tables and queues that can be accessed by REST through HTTP or HTTPS.

The vital entities of this CSP are Azure database, data sync and access control component. The Azure database is based on Microsoft SQL Server and managed by a special Database Management System (DBMS). The data access is performed by the ActiveX Data Object (ADO.NET) and other component data access interfaces of .NET framework. The 'data sync' component synchronizes the stored distributed data.

The access control component checks the characteristics of the cloud user for providing access to the cloud data. The data access is done by advertising the service endpoints, through which the services are invoked by the fellow applications and all the service endpoints are detected with a unique Uniform Resource Identifier (URI).

1.11.6 Google App Engine (GAE)

GAE is very famous when compared to the standard web services, as it supports the programming languages such as Java and python. In GAE, the cloud users are allowed to upload the data to the cloud server and the GAE performs the data storage and retrieval through a BigTable.

This BigTable is a non-relational database and the developers are given the read-only permission for the data present in GAE. The merits of GAE are self load balancing and scale-up features.

Thus, this section presents the most common CSPs in the current market and the research motivational ideas are presented in the following section.

1.12 MOTIVATION OF THE RESEARCH

Cloud computing is an evergreen technology that supports the business community with a wide range of services. Based on the business requirements, the client end chooses one or more services from the provided services. Even though the cloud computing enjoys numerous benefits, it still suffers from certain crucial challenges.

The most serious challenges faced by cloud computing are data integrity and availability. Data integrity is the most important requirement, which brings the trust on the CSP. Failing to achieve data integrity leads to severe problems such as data leakage, data tamper, sniffing and so on. On the other hand, the cloud follows distributed architecture such that the data availability is hard to achieve.

This research work attempts to combat against both these issues by decomposing the research work into three phases. All the phases focus on ensuring data integrity, whereas the second and third phases are meant for supporting data availability.

1.13 RESEARCH OBJECTIVES

The overall goal of this research work is to ensure data integrity and reasonable availability of cloud data to the cloud users. The data integrity of the data is ensured by strict security based algorithms and the availability of the cloud data is promised by incorporating data management over multiple cloud servers and replica creation strategies. This research work sets the objective by considering two solid problems and the objective is met by presenting three different research solutions. The specific research objectives are listed as follows.

- To present a twin layered security scheme for cloud storage for ensuring data integrity.
- To propose a secure storage model for cloud computing environment by employing a distribution agent.
- To introduce a secure data replication management scheme with better data accessibility.

When the research objectives are fixed, the research problem is fixed and broken down into three research phases. The research problem formation is presented in the following section.

1.14 RESEARCH PROBLEM STATEMENT

The research problem statement is formed after fixing the objectives of the research work. The complete research work is decomposed into three phases a and the summary of all the research phases are presented as follows.

1.14.1 1 Research Phase I

The main aim of this research phase is to present a twin layered security mechanism, which preserves the data integrity of cloud storage. The proposed work verifies the data integrity of the cloud stored data by employing the keccak's hash function and Advanced Encryption Standard (AES) for carrying out the encryption operation.

Thus, the proposed work provides twin layered security. The performance of the proposed work is analysed against the Secure Hash Algorithm 3 (SHA-3) member algorithms such as BLAKE, SKEIN and SHA2 with respect to execution speed and memory consumption. Besides this, the time consumption for encryption is also presented. The highlights of this research phase are listed as follows.

- Twin layered security mechanism is presented for the cloud storage, which involves hashing and encryption.
- Data nodes organized in structured tree format allows updation in any part of data at any instant of time.
- No table look up is required and no tables are maintained. This reduces the memory consumption and improves the performance.
- Keccak's hash function is employed for hashing, as the degree of security and the processing speed are considerably improved.
- The proposed work is free from any arithmetic and rotation operations.
- AES algorithm is utilized for encryption, as it is known for its speed and efficiency.
- On analysing the performance of the proposed work, it shows surprising results when compared to the analogous techniques.

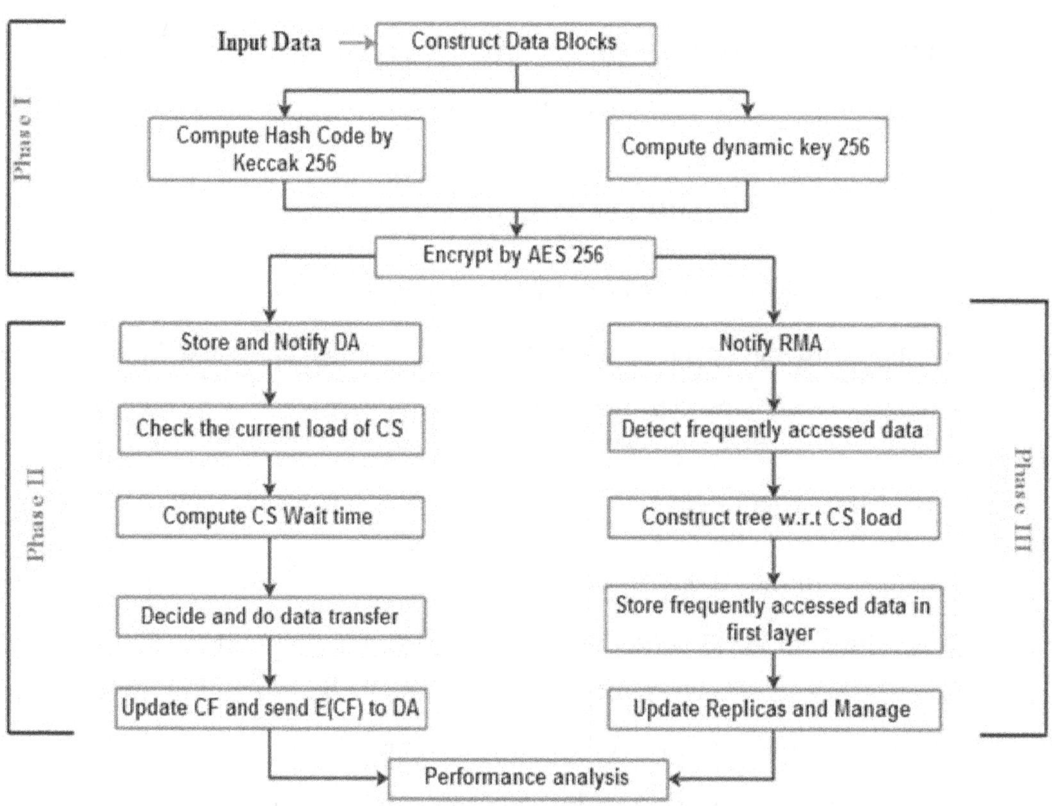

Figure 1.13: Overall Flow of the Research Work

In substance, this phase tightens the security by introducing two layers of security, which are hashing and encryption techniques. The data are organized in the form of tree, which makes it easy to perform operations such as data update, insert and delete. The experimental results of the proposed work are compared with the member algorithms of SHA-3 with respect to throughput, memory consumption and execution time. The overall flow of the research is depicted in Figure 1.13.

1.14.2 Research Phase II

In continuation to the previous research phase, the data block separation is performed by Keccak's hash function and the encryption is carried out by AES algorithm. The encrypted data blocks are distributed by the agent to different cloud servers by taking the current load and traffic of the server into account.

The flow of the data is marked in a separate file called catalogue file, which is maintained by the agent. This file is shared with the data owner, followed by which the catalogue file is deleted, in order to conserve memory. This provides a strong security as the proposed work sets several checkmates for the intruders. The highlighting points of this phase are listed below.

- This work separates the entire data into several data blocks, which is followed by the encryption phase. The encrypted data blocks are distributed across multiple cloud servers in a random fashion.
- The choice of cloud server depends on the current data load and traffic of the cloud server.
- A special file named catalogue file is maintained by the agent, which keeps track of the sequence of the data.
- The catalogue file is shared with the data owner and deleted permanently by the agent. This saves memory and reduces the computational complexity.
- The catalogue file is encrypted by lightweight symmetric key encryption.
- This work is faster, as the encryption is performed in a parallel fashion.
- As the owner is well-informed about the exact location of all data blocks, the data owner can check or modify that particular data block. This reduces the computational overhead and speeds up the retrieval process.

The highlighting points of the third research phase are presented below.

1.14.3 Research Phase III

In order to achieve data availability, the cloud data is replicated and stored. The replicas increase the data availability, however introduces several challenges such as data modification and deletion. When the original cloud data has undergone some modifications, the same modifications have to be reflected on the replicas also, which is highly challenging.

The modifications must reflect the replicas immediately, so as to escape from data inconsistency. These prompt modifications are possible only when the replicas are organised by some means. Considering these facts, this work attempts to organize the replicas in tree format by taking the load of the Cloud Server (CS) into account. The highlights of this phase are as follows.

- The replicas are not created for all the data, but for data with high access frequency. This idea saves space and time.
- The created replicas are organized by means of a tree like structure, which is formed on considering the CS load.
- The frequently accessed cloud data are stored in the first layer of the tree, which helps in easy localization of the data.

Hence, this section presents the highlighting points of the research and the following section discusses about the organization of the thesis.

1.15 THESIS ORGANIZATION

The whole thesis is categorized into seven chapters and the outline of each and every chapter is presented below.

Chapter 1 presents the fundamentals of cloud computing with its functionalities and challenges. In addition to this, the motivation and the objectives of the research are also presented.

Chapter 2 reviews the security facets of cloud and the related literature with respect to cloud data security and availability .

Chapter 3 presents a twin layered security model, which guarantees the security of the data and thus the data integrity is preserved. The proposed work involves two building blocks, which are hashing and encryption. The input data is separated into several blocks, followed by which a tree is constructed. This is followed by the computation of hash codes, which is done by the keccak-256 hash function. The hash codes are then encrypted by AES

(256) with a dynamic key being generated by CodeIgniter. Thus, the security of the system is tightened and misuse of the data is strictly denied. The performance of the proposed work is tested with respect to throughput, memory and time consumption.

Chapter 4 proposes a Secured Storage Model for the cloud computing environment, which employs many parameters to ensure data integrity. Initially, the original data is divided into several data blocks and encrypted by AES algorithm, for which our previous work Twin layered Security Scheme (TSS) is utilized. On successful completion of encrypting the entire data, the encrypted data blocks are distributed through the cloud servers with the supervision of the Distribution Agent (DA). The DA selects the cloud server by taking the current load and traffic of the cloud server into account. Additionally, the DA is responsible for managing a special file named catalogue file, which keeps track of the location of the encrypted data block. The catalogue file is permanently deleted by DA, once it is successfully shared with the data owner. The proposed work places three strong checkmates to the intruder and hence, it is very difficult for the intruder to break the checkmates and pass through.

Chapter 5 introduces a Replication Management Scheme, which creates replicas by considering the access frequency of the data and the replicas are loaded onto the cloud server by considering the current load of it. This idea balances the load of the cloud server. All the replicas are organized in a tree like structure and the replicas with maximum hit ratio are placed on the first level of the tree to ensure better data accessibility. The performance of the work is satisfactory in terms of data accessibility, storage exploitation, replica allocation and retrieval time.

Chapter 6 presents the conclusive points of the research which presents both the sides of the coin. Possible future enhancements are also listed in this chapter.

CHAPTER 2

REVIEW OF LITERATURE

Cloud computing is the upcoming technology because of its performance, minimized implementation and maintenance costs. Increase in the mobility for global workforce, infrastructures supporting scalability, quick time to market, transformation in IT departments, high performance applications to business, greening of data center are the benefits of cloud computing.

This section outlines the basics of cloud data security in section 2.1 and the section 2.2 discusses about the security issues in SaaS. Section 2.3 explains the basic concepts of cloud security and related works are discussed in section 2.4.

2.1 BASICS OF CLOUD DATA SECURITY

Security is defined as the condition of an entity to be secure and can sustain security threats. Information security is defined by Committee on National Security Systems (CNSS) as the information safeguard, which involves all the systems and hardware being employed for data storage and transmission. As the data volume is growing every single data, confidentiality is the most important concern of today's data.

Computer security is defined as safeguarding the information from the users without having official permission to do something .Computer security is the process of securing and preventing the resources from unwanted users to accomplish the intention of preserving the integrity, availability and confidentiality . The cloud resources consider all sorts of resources such as hardware, software, firmware, data and networking equipments.

A worldwide survey indicated that about 28% of data demands secure storage out of 1.2 Zetta Bytes in 2010. By the end of 2015, the data to be stored securely raised to 33%, as cited by Krutz and Vines [2010]. Hence, information security is one of the serious threats to the cloud computing environment.

When information security is considered, indirectly it refers to two other aspects of security such as network and data security. The security of cloud computing is based on several layers involved in it and the important security aspects are discussed as follows.

- Physical security - This kind of security is required for the physical devices such that misuse of physical components can be avoided.

- User security - User security is meant for safeguarding the organizational data being associated with the user.
- Network security - Communication security must be ensured for safeguarding the data during data communication or forwarding.
- Information security - Information security is concerned with multiple facets of security such as data integrity, confidentiality, privacy, availability and so on.

In addition to this, the security threats for cloud computing can be categorized on the basis of the cloud service modes such as SaaS, PaaS and IaaS. The security issues faced by SaaS are presented in the following section.

2.2 SECURITY ISSUES IN SAAS

As mentioned in the previous chapter, the cloud service user relies on the CSP for ensuring security. The CSP highlights the security provisioning ability in the SLA itself. However in certain cases, the CSP may be unable to meet the security requirements. Subashini and Kavitha [2011] claimed that as the cloud computing supports multi-tenancy, the data must be safeguarded from the other users. This makes it difficult to provide better security to the cloud consumers.

The SaaS software provider deploys the software application on a private server or deploys it on a cloud computing infrastructure service provided by a third party provider. Most of the today's organizations consider the data and business processes as treasure and safeguard them with strict access control and other security based strategies. However, in case of SaaS, the organizational information is stored at the data center of the SaaS provider, in addition to the information related to some other organization.

In order to boost up the data availability, the CSP may replicate the data and store them at different locations. Though this idea seems to be beneficial, several security breaches are introduced by data replication. Some of the important security elements meant for SaaS applications are depicted in the following Figure 2.1.

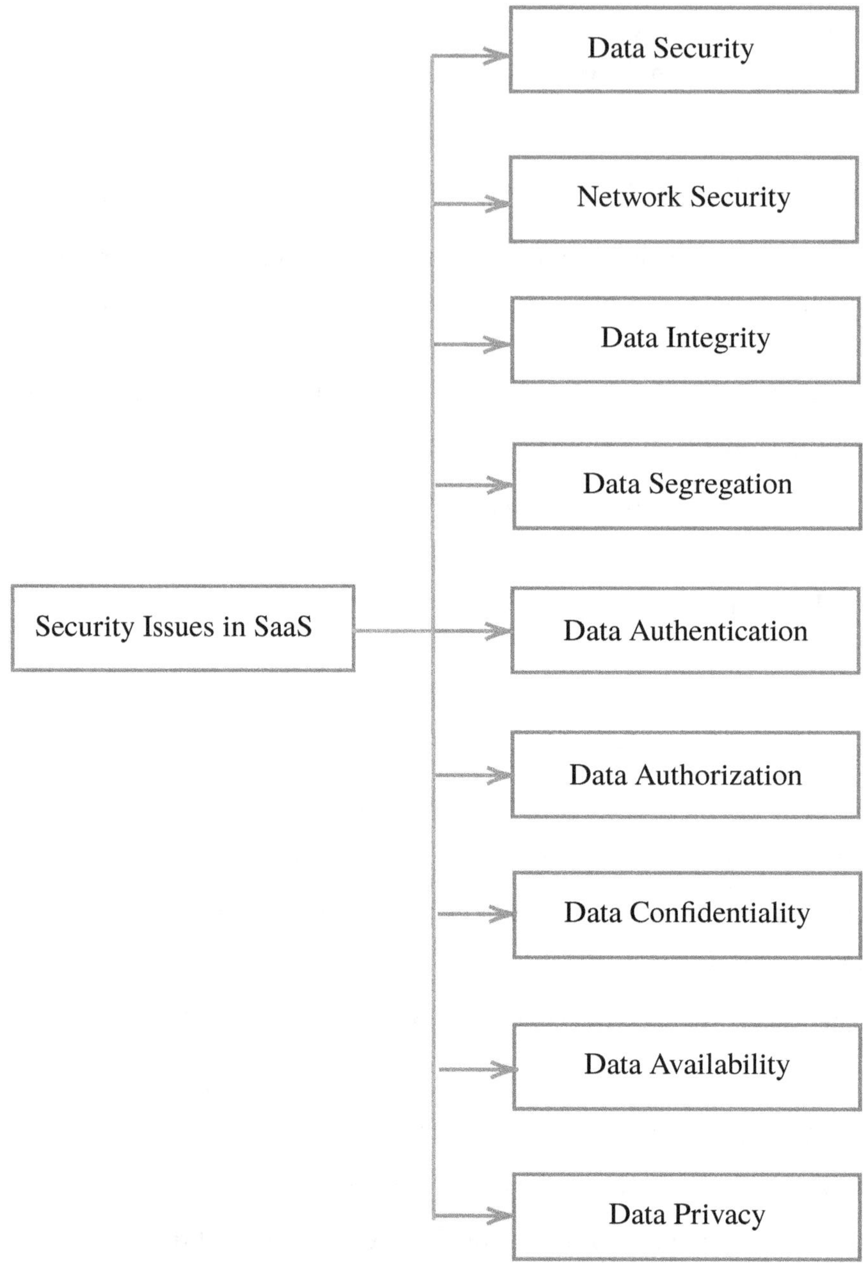

Figure 2.1: Security Problems Faced by Saas

All these cited security problems of SaaS are discussed in the following sub-sections.

2.2.1 Data Security

Cloud computing environment prompts the users to outsource the user's information on an external storage space provided by a third party. When the data is contracted out to the external server, then the data owner feels like the data control is completely lost and the data may be stolen by malicious attackers. The malicious attackers can be either internal or external. Certain cloud servers may have a simple access control policy, which in turn allows malicious attackers to gain unauthorized access to the data. The NIST listed data loss and leakage is one among the major threats in cloud computing environment. As in Zhou *et al.,* [2012], Ahmed *et al.,* [2010], Gul *et al.,* [2011], Sen [2015], Bhadauria *et al.,* [2011], Sun *et al.,* [2011] information loss and leakage happens because of the employment of poor security policy in storage and transmission. This problem can be addressed by considering different aspects of security as pointed out below.

- **Data during transmit:** Data during transmit indicates the data at the time of data transmission from and to the cloud server.
- **Still Data :** Still data is the data that rests in the cloud server without involving any data based operation.
- **Data line:** Data line represents the complete information about the data right from the utilization by applications and the data utilization for auditors. However, achieving data line is difficult for public clouds.
- **Data provenance:** Data provenance does not concern with the data integrity alone but the complete life cycle of the data which includes data creation, alteration or deletion of data stored in the cloud server.
- **Data remanence:** Data remanence indicates the remaining data after the process of data deletion.

The traditional application deployment scheme within an organization intends to identify the confidential data and stores it in memory by imposing strict security policies. However, in the SaaS model, the organizational data is outsourced to the cloud. Hence, the SaaS provider is responsible for ensuring security guarantees by avoiding security breaches. This is possible by employing stronger encryption based techniques in addition to authorization.

Basically, the popular CSPs such as Amazon do not access the user instance and cannot pass through the guest operating system. All the data accesses made are registered properly and regular auditing is performed. However, when the data is stored in Amazon S3, the data is not encrypted by the CSP, however the data can be encrypted on the user side before data upload. This makes sure that the data is not modified by malicious users at any cost. Certain tests are carried out by the SaaS vendor for providing the data security owned by an organization and the tests are listed as follows.

- Cross site scripting (XSS)
- Access control weaknesses

- Cross Site Request Forgery (CSRF)
- Cookie manipulation
- Hidden field manipulation
- Insecure storage
- Insecure configuration

The illegitimate accesses to the data can be prevented with the help of these tests and the following section explains about the network security.

2.2.2 Network Security

Data loss and leakage is claimed as the second most popular threat by NIST. Information loss and leakage happen because of the poor security policy during data storage and transmission. Usually, confidential data is collected from the organization and some processes are carried out by the SaaS application, followed by which the data is stored at the SaaS provider site. Every communication activity involves data transmission, which may get leaked during data transmission.

In order to secure the data during transmission, the traffic encryption approaches based on Secure Socket Layer (SSL) and the Transport Layer Security (TLS) are utilized. In Amazon Web Services (AWS), the network layer offers better security in opposition to standard network security issues such as man in the middle attack, IP spoofing, packet sniffing and so on. Amazon S3 ensures better security as it can be accessed through the SSL.

The encrypted endpoints can be utilized from either the Internet or the Amazon EC2. In spite of this, the illegitimate users may sniff the drawbacks in network security configuration to intrude the network messages. Certain evaluation tests for checking the network security of the SaaS vendor are as follows.

- Network penetration and packet investigation
- Session management weaknesses
- Insecure SSL trust formation

Failing to satisfy these checkmates paves way for unwanted illegitimate activities and the following section explains about the core concept of this research, which is data integrity.

2.2.3 Data Integrity

Data integrity is a crucial issue confronted by cloud computing environment. The data integrity ensures that the data is not altered or modified by an illegitimate user in any means. Hence, the data integrity guarantees that the data is original, complete and protected from malicious users. Data integrity refers to protection of data from unauthorized modification, insertion and deletion. Zissis and Lekkas [2012] stated that the data can be crooked as the cloud computing environment, as it supports resource sharing. So data integrity is also a most crucial requirement of any data processing system and it can be attained in a standalone system effectively. However, data integrity is difficult to achieve in distributed environment. Integrity is one of the important ingredients of the ACID properties of Database Management System (DBMS). The integrity is represented by the alphabet 'I' of ACID property.

This property can be managed by enforcing constraints. Most of the databases strive hard to provide ACID properties. Digital Signature is a popular mean to protect the data integrity and an effortless way to provide data integrity is Message Authentication Code (MAC). The simple way for providing integrity is using message authentication code (MAC). MAC is stated as a cryptographic checksum being computed by hash functions, which is forwarded in addition to the data for performing integrity check. These auditing mechanisms are also used for integrity preservation. In case of private auditing, the data integrity is evaluated by the data owner with certain algorithms. On the other hand, the public auditing nominates a Third Party Auditor (TPA) for analysing the data integrity. This TPA does not have the rights to access the data but it checks for the data modification. However, in case of distributed environment, transactions are needed to be handled with respect to numerous data sources and are achieved by the global transaction manager.

For this sake, a two phase commit protocol based on XA standard is utilized. Every SaaS based application is varied in terms of availability and SLAs, hence ensuring data integrity is bigger challenge. When the data integrity is compromised, then serious issues arise. Hence, the developers focus more on data integrity when it comes to cloud architecture.

2.2.4 Data Segregation

As mentioned earlier, multi-tenancy, sharing of resources is the greatest benefit of cloud computing and hence numerous users store the data in cloud. This implies the data of multiple users are accommodated in a same place. This may result in the unwanted activity of intrusion through the data of other user.

The intrusion may be carried out by hacking the application or by including client code into the SaaS application. The illegitimate user may arrive at a secured code and may include the code into the application. Suppose when the SaaS application is executed without any security checks, then intrusion happens and is highly unfavourable.

This issue can be treated properly, when the SaaS model places a clear-cut segregator between the data of every single user. The data segregator should be present at both the application and physical level. Additionally, service should also be equipped such that it can differentiate the data obtained the users.

The illegitimate user intends to pass through the security checks by utilizing application susceptibilities for accessing the confidential data of other users. Some of the assessment tests meant for analysing the data segregation are the flaws in SQL injection, data validation and unprotected data storage.

2.2.5 Data Authentication and Authorization

Authentication is a protection mechanism which ensures that the data is given access only to the authorized users. Authentication is the process of confirming the identity of a user. It is mandatory to have strong authentication mechanism for cloud deployment. Authentication is a basic security policy to ensure access control. When a user submits the access credentials, then the credentials are authenticated before provisioning the access to the data. Hence, authorization controls the data access by the users and it permits the correctly recognized users to use the resources.

Mostly, the details about employees are stored in the Lightweight Directory Access Protocol (LDAP) servers. The software is deployed, which is not under the control of the corporate firewall in SaaS and the user login credentials are saved in the databases provided by SaaS and organization's database is not utilized for storage.

Hence, the database in SaaS must be updated then and there, such that the new employees can be included and the old employees can be removed from the list. SaaS providers prompt the users to pass through the process of authentication by employing LDAP servers, such that organizations can preserve the control over the users.

2.2.6 Data Confidentiality

Data confidentiality is very essential for the storage of confidential or private data. Certain data are private and are needed to be kept confidential. Data confidentiality is achieved by strong authentication and access control policies. It is not feasible to store the private data as such, in the cloud storage. The information is stored rather in unintelligible format, such that unintended users may not infer any knowledge from the stored data.

Encryption is also used to preserve data confidentiality. The process of intends to convert the plaintext into ciphertext. Encryption encodes the data from human readable form such that they can be decoded only by the authorized users having the key. Cryptographic algorithms are used for encryption.

However, encryption of cloud storage introduces several complexities to the system such as query based search, authorization and data modification. Some of the methods to achieve data confidentiality are homomorphic encryption, encrypted search and database, distributive storage and the hybrid of the techniques.

a Encryption

Rivest *et al.,* [1978], Ryan [2013] introduced homomorphic encryption . This type of encryption is simple, because it generates steady decryption results and data modification doesnt need any data decryption. Gentry and Boneh [2009] presented a complete homomorphic encryption, which allows all sorts of operations over stored data without the process of decryption.

However, encryption is not the best choice to assure data confidentiality. This is because of the increased complexity and the cost it consumes for storage and computation. Kaur and Bhardwaj [2012] propose a hybrid approach which clubs several encryption algorithms such as RSA, Triple DES and pseudorandom number generator is presented to provide security.

b Encrypted Search and Database

Homomorphic encryption is proved to be ineffective and thus the researchers started to pay attention towards encrypted search. Manivannan and Sujarani [2010] proposed a lightweight mechanism to encrypt the data by means of transposition, substitution, folding and shifting operation . Yet, the number of keys is directly proportional to the processing time and cost.

Huang and Tso [2012] proposed an asymmetric encryption based technique . In this work, commutative encryption is applied several times over the data and re-encryption is made possible such that the cipher text is encrypted in a repetitive fashion. A privacy preserving multi-keyword ranked search over encrypted data is presented by Cao *et al.,* [2013].

c Distributive Storage

This distributive storage is another option to provide security in cloud computing. The security related problems with respect to data privacy that focuses on data integrity, intrusion and data availability is presented by AlZain *et al.,* [2011]. The work proposed by Ram and Sreenivaasan [2010] introduced security as a service for cloud computing environment. This technique decomposes the cloud data into multiple pieces of data. These pieces of data are then encrypted and distributed among several databases. This technique offers guaranteed security.

d Hybrid Methodology

The technique proposed by Rao [2012] exploits both the key sharing and authentication methods to provide data confidentiality and integrity. A data security method, which relies on three layers is proposed by Mohamed *et al.,* [2012]. The first layer ensures the authenticity of the user, the second layer encrypts the data and the last layer is responsible for faster decryption process.

e Deletion of Data

The data must not be available to the users, when the data is deleted after confirmation. This is not easy to achieve because there may be multiple copies of the same data in the cloud environment. When the users delete the data, all the replicas must be deleted all at once.

f Challenges in Attaining Confidentiality

The major task of cloud computing is the data outsourcing, at which the users store the data in the remote servers provided by the third party vendors. These outsourced data can be accessed from the remote servers via internet by proving the identity.

The original data of the users can be stored on a single or multiple cloud providers. In this case, certain confidentiality issues may arise and some of them are as follows.

- Whether the cloud computing environment supports the data privacy and confidentiality for data storage.
- Whether the user is enlightened with the associated privacy and confidentiality risks with respect to the data storage in addition to the policies.
- Data disclosure and isolated data storage may end up in unfavourable consequences such that the original legal status of data protections for personal and business information are managed.
- The privacy and confidentiality of the data may be affected by the location of information stored in the cloud.
- Data in cloud storage may present in more than a single legal location at a given time with different legal effects.
- Sometimes the laws necessitate a CSP to investigate the user records for substantiating the criminal activity and other security related issues.
- Legal doubts make the process difficult for fixing the current status of information in the cloud in addition to the privacy and confidentiality protections available to users. Hence, the important points about data confidentiality are discussed.

2.2.7 Data Availability

Data availability is the significant feature of cloud computing that makes the service available for the users at all times. The data availability can be achieved by better data scalability and availability. In order to achieve better data availability, a multi-tier architecture is utilized and supported by a load balanced application instances that can be executed on several servers.

The application must sustain the hardware and software failures and other forms of security attacks, such that the data availability is ensured at all times. In case of any unfavourable situation, the data must be able to get restored and utilized. This is very important for the safety of the organization's data and also a reasonable availability ends up in better data accessibility.

For instance, in case of the Amazon Web Services (AWS) the endpoints are hosted on an Internet-scale and world-class infrastructure such that it can support the Amazon.com retail site. The Standard Distributed Denial of Service (DDoS) attack avoidance techniques such as synchronous cookies and connection limiting are employed to ensure protection. The main causes that could damage the availability of the data are authentication weaknesses and session management weaknesses.

2.2.8 Data Privacy

Privacy is the ability of an individual or group to seclude themselves or information about themselves and thereby reveal them selectively Mitali and Sharma [2014]. Privacy is the basic human rights and needed to be preserved.

- **Service exploitation -** The data stored in the cloud may suffer from data leakage, due to negative service exploitation. Thus, proof of ownership is claimed by the cloud to preserve the sensitive data from unwanted exploitation.

- **Privacy attacks -** Privacy attacks can be overcome by the trustworthy cloud computing. The trusted model must provide data confidentiality. There are several methods proposed by having trust as the underlying platform.

- **Identity management -** Involvement of TTP increases the security threat. This problem can be solved by the exploitation of TTP. There are several degrees of protection mechanisms, so as to prevent data leakage and privacy loss. Trust based architecture upgrades the cloud services by providing security.

Hence, this section discusses the major security facets of cloud computing and the short notes of every security issue of the cloud.

2.3 SECURITY ISSUES IN PAAS

With PaaS the cloud users are provided with a range of control for building applications in the cloud platform by the CSP. However, the security beneath the application level, for instance network intrusion prevention is under the service providers control, such that the provider has to provide strict guarantees that between the applications the data stays unreachable.

Paas is meant to allow the developers for building own applications on the cloud platform. Thus, Pass is more extensible when compared with SaaS and that too at the cost of customer-ready features. This difference reflects in the features of security and capabilities where the predefined capabilities are minimal. In addition to this, more flexibility is available to the layer on additional security.

Web Service (WS) security protocol is provided for protecting the Enterprise Service Bus (ESB) in applications which are adequately difficult to influence an ESB. The capability of segmenting ESBs is unavailable in PaaS environments. To quantify the efficiency of the security procedures involved, metrics must be properly in place.

Vulnerability scores and path coverage are considered as security related metrics with respect to the direct applications. Both these metrics represent the quality of the application coding. The application must consider the activities of malicious users on a new cloud application.

Hackers may tend to corrupt the visible code, which could include the code executed in the user environment. They could corrupt the infrastructure and carries out general functional testing or black box testing. The security threats present in the cloud are related not only to the web applications but also the security threats related to the machine-to-machine Service Oriented Architecture (SOA) applications. These kinds of applications are nowadays increasingly being deployed in the cloud. On a general basis, the important threats to be handled by the PaaS are as follows.

- Default application formation
- SSL protocol, code execution flaws and insecure cloud data permissions
- Authentication, Access Control and Authorization (AAA)

With the presence of these risks and security threats, the issues stated above directly affects the cloud environment. The SaaS threats may also affect the PaaS and should be focused as well, as claimed by Subashini and Kavitha [2011].

2.4 SECURITY ISSUES IN IAAS

In the case of IaaS, the developer is provided with a direct control over the security when the virtualization manager is strong enough. However Attanasio [1973], Gajek *et al.,* [2007] presents the security issues of virtual machines theoretically rather than practically. The second important factor is the data reliability as the data is stored in the hardware of the provider.

Owing to the advancement of virtualization technology, data control is the important requirement irrespective of the data location. Descher *et al.,* [2009] claims that the trust and security of a resource in the cloud can be ensured by employing several techniques.

The security requirements of the consumer and the provider differ from each other and it depends on the cloud service models as well. For instance, as described by Amazon [2009],

IaaS includes the vendor responsibililty for security up to the hypervisor which means that the EC2 could address security controls such as physical, environmental and virtualization security.

Seccombe *et al.,* [2009] denotes that the user is accountable for the security controls that are associated with the IT system which includes the OS, data and the applications. Hence the cloud architecture must be able to provide better security and user reliability. All organization and control communications between CSPs and users should be carried over the secure channels employing protcols which are based on standards so that better features of security related to authentication, authorization, data confidentiality, data integrity and data accountability could be provided. The absence of these security measures could lead to data leakage, unauthorized access, identity misuse and so on.

Hence, all these security measures must be provided by the CSPs. It would be better if the important processes such as signup, resource provisioning, money payment, monitoring of the resources and other functions related to management and control is made simpler. Hence the service providers must offer strong mutual authentication schemes for the service users by which the user accounts are offered a better security.

In case, when a service provider makes use of password authentication scheme for rendering access to the user account, functions related to control and management, it becomes the responsibility of the provider to ensure that the chosen passwords are strong enough to break and they are passed through encrypted channels alone. Nowadays a minimal number of service providers impose strong password formation rules. Further, access monitoring for identifying and preventing deceptive access is also necessary.

Usually, the resources allotted by the service provider do not communicate among themselves unless they are owned by the same user. The term 'resources' includes the computing, resources involved in networking and storage purpose, applications, services and other related objects. The user to which a resource is allotted is called as the 'owner'.

This concept makes sense that the objects present in the cloud environment are not unintentionally uncovered to users who are not authorized. The provider has a choice of providing the users a mechanism to provide or grant access to their resources and allow other customers with a right to use their data and services.

A denial policy as default is suggested to be considered irrespective of the access configurations which are possible. When a part of the resource is wanted to be pooled or shared between multiple customers, it is necessary that the provider is providing some security controls.

The provider should impose adequate protection to avoid the access of data and services by unauthorized users in an unauthorized way, modifying or destructing the objects. This means that it should be able for the provider to express that the security provisions are intact by avoiding the user A from accessing, changing or destroying the resources lined up with the user B. Usually, the controls are validated with the help of a trusted third party or otherwise an auditor. Further the provider should be able to provide the audit reports to the users based on their requisition.

To avoid the accidental or the malicious access of the resources, the service providers must be able to apply controls over the resources which are present under their control. Since the providers are able to access the infrastructure physically, many general security issues can be avoided.

Sometimes it happens that the providers use their controls such that the own employees could not access their own resources. It is mandatory hat all the legitimate accesses must be reviewed and audited periodically for promoting.

Hence, the security issues of IaaS, PaaS and SaaS are discussed in this section.

2.5 CLOUD SECURITY

In spite of numerous benefits, certain security requirements prevent users from utilizing cloud computing. With the emergence of cloud computing, the cloud users are allowed to store data on a remote storage location which is managed by a third party. When the data is under the control of a third party the owner can no longer have control over data and the data is susceptible to damage or loss. The one who attacks the data may be internal or external. Due to the weaker access controls, unauthorized access is more probable .By storing the data on remote server user lacks control over the data which leads to security problems such as data integrity, data confidentially and data availability. To overcome these problems mechanisms are adopted by cloud users to make sure that data was intact in the remote server and are accessible whenever needed. Security and auditing mechanisms are adopted by cloud for ensuring data confidentiality and integrity.

Following are the challenges that arise for the protection of data.

- Access control: It is a mandate thing to check if there are appropriate controls over the access of information.

- Structured versus unstructured: It deals with the storage of data and the ease of access provided.
- Integrity/availability/confidentiality: How the triads of information security are maintained in the cloud?
- Encryption: Some information is stored only in the encrypted form. It is the duty of the cloud user to see if such requirements are satisfied by the CSP.

Rong *et al.,* [2013*a*] presents the major security issues in data storage are: data confidentiality, data integrity and data availability.

2.5.1 Confidentiality

One of the major dispute in cloud computing is confidentiality. Confidentiality relates to the secrecy of the data. It is strongly preferred to be accessed by the authorized users alone. Revealing the original data leads to the loss of confidentiality.

Storing data in remote servers arises questions regarding the confidentiality of data.

- Will the data stored be secret?
- Will the CSP be honest?

To ensure the confidentiality and availability, the commonly used efficient mechanisms are encryption algorithms and strong authentication techniques. The process of converting the data in the form of plaintext to cipher text is called as encryption. Keys are used to encrypt the data. When the key is lost, the aim of encryption gets lost. The major concerns regarding encryption are:

- How is protection given to the data in the cloud?
- What is the key strength of the encryption?

Rahmani *et al.,* [2013], Rao and Selvamani [2015] states that the encryption choice depends on the CSP. CSP normally encrypts the data prior storing the data and the keys are disclosed only to the authorized persons. There are different cryptographic algorithms for doing encryption. The encryption algorithms are either symmetric or asymmetric. Symmetric algorithms use private key for both encrypting and decrypting the data. DES, AES, Blowfish are some of the common symmetric algorithms.

Mitali and Sharma [2014] states DES is a popular symmetric key encryption algorithm. It was introduced in 1976 and widely used in many applications both commercial and financial.

It is easier to implement in terms of hardware and software. Its poor performance is a drawback. AES encryption algorithm slowly replaced DES. It proved to be faster and more flexible when compared with the DES. It is used to give protection for data in smart cards and online transactions. AES has a 256 bit key and it is more secure. Inspite of its advantages, AES can be complex too. Blowfish is another most commonly used public domain encryption algorithm. It is introduced in the year 1993. It is a fast and simple encryption algorithm.

Symmetric algorithms are generally faster and simpler but not efficient up to the mark as the sender and receiver share the same private key. The second category of encryption algorithms is asymmetric algorithms otherwise called as public key encryption algorithms. It involves the use of public key and private key. The data is encrypted with the public key of the receiver in the sender side and the receiver will decrypt the message with his private key.

One of the most famous asymmetric encryption algorithms is RSA encryption algorithm. It is developed in 1978. It provides an increased security to the data when compared to the previously used algorithms. It provides mechanisms for applying digital signatures. It is a mathematical scheme for proving the authenticity of data. Predicate encryption is a kind of asymmetric encryption which allows decryption of selected data instead of decrypting all the data.

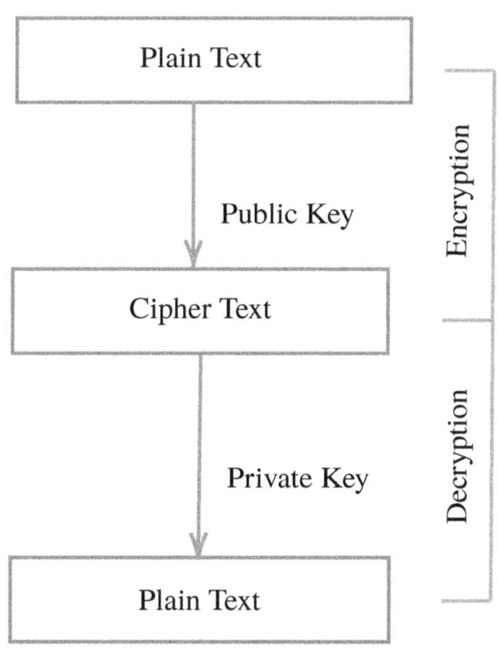

Figure 2.2: Asymmetric Encryption

Identity Based Encryption (IBE) is a public key encryption which uses the unique information about the identity of the user as public key and guarantees authenticity. When symmetric algorithms and asymmetric algorithms are compared with each other, symmetric algorithms are faster than the asymmetric ones. In terms of security, asymmetric algorithms are better than the symmetric algorithms. Figure 2.2 depicts the asymmetric encryption technique.

Han *et al.,* [2014] proposed a work on searchable encryption. The algorithms earlier used are all under the category of encrypting and decrypting the entire file. It is a time consuming task. As an alternative, searchable encryption is introduced. It allows to build an index for the file with the keywords and the file is encrypted and stored. While searching the data, only the keywords are decrypted rather than the entire file and the search is made.

Another efficient encryption algorithm is the homomorphic encryption in which operations are carried out on the encrypted file by the CSP which has the same results as the decryption method. As stated by Tebaa and Hajji [2014], the key used for encryption is kept secret by the user and not revealed to the CSP. This results in more secure encryption. In other encryption algorithms, maintaining the secrecy of the key used for encryption is a challenging task for the CSP as the users are growing day by day and lot of data is dumped into the cloud. Moreover there is a chance for the CSP to hack the data. So the presence of key only in the hands of the user improves the security of the data.

2.5.2 Integrity

Integrity is a serious problem faced by the cloud computing. It is ensuring that the data has not been changed or altered by any unauthorized person. It is a method for making sure that the data is genuine, accurate and protected from unauthorized access. Resource sharing being the major characteristic of cloud computing, there is a great possibility for the data to be corrupted by unauthorized users.

To preserve the integrity of data, digital signatures are used. The integrity of data is preserved in a simple way with the use of Message Authentication Code (MAC). It is a cryptographic checksum calculated using hash functions. This checksum is sent along with the data being sent to check the integrity of data. For this purpose of preserving integrity of data, auditing mechanisms are also used.

Auditing may be private or public. In private auditing, the integrity of the data is checked by the data owner using algorithms. The public auditing means the data integrity is checked by a Trusted Third Party (TPA). The TPA can have no access over the data, but can verify if the data is been modified or not and reports it to the owner of the data.

Remote data auditing refers to a group of protocols for verifying the correctness of the data in the cloud managed by the CSP without accessing the data. The challenge response process is used in the remote auditing. It involves the following steps.

- The file is processed by the data owner. Metadata is generated and it is handed over to the TPA.
- A challenge is generated by the TPA and it is transmitted to the CSP to check the correctness of the data.
- The CSP calculates the response and sends it to the TPA on receiving the challenge.
- Once the TPA finishes the verification of the response received, the data is stored correctly by the provider as cited by Rong *et al.,* [2013*b*].

Provable Data Posession (PDP) is also a remote auditing mechanism in which either the data owner of the TPA will check the integrity.TPA is not able to check the integrity independently and needs the help of the data owner. Only when the data owner sends the metadata for verification, the TPA exactly verifies. Without the knowledge of the data owner, the TPA does not have any permission to take countermeasures. This issue is overcome by introducing a proxy PDP which is assigned the remote data auditing task as described by Wang [2012].

2.6 BASIC CRYPTOGRAPHIC ALGORITHMS FOR CLOUD SECURITY

In order to provide secure communication over the network, encryption algorithm plays a significant role. It is the fundamental tool for protecting the data. Encryption algorithm transforms the data into scrambled form by using "the key" and the user alone has the key to attain data decryption. In Symmetric key encryption, only one key is used for encryption and decryption. On the other hand, in asymmetric key encryption, two keys- private and public keys are utilized. Public key is used for encryption and private key is used for decryption. The division of cryptographic algorithm is depicted in Figure 2.3. The basics of the algorithms are presented as follows.

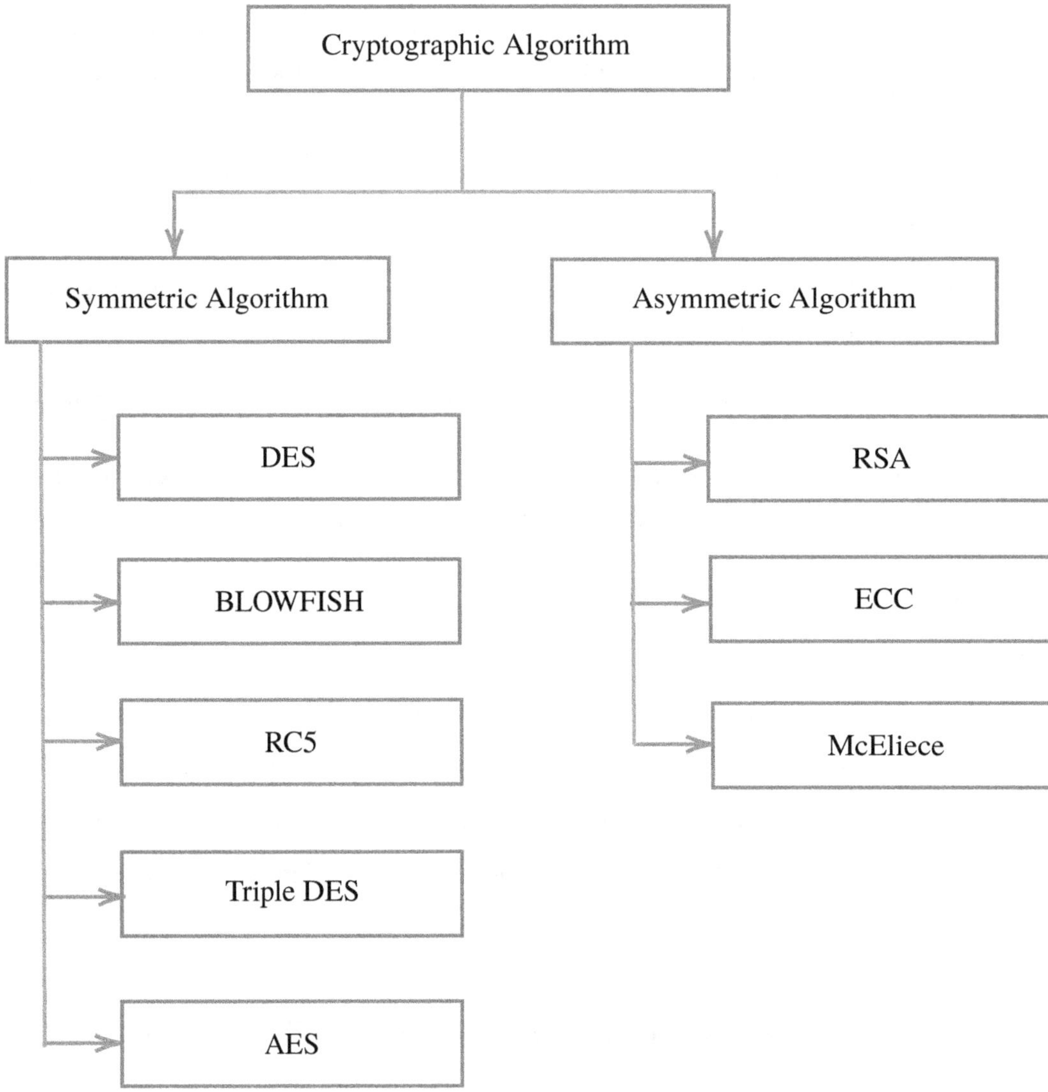

Figure 2.3: Cryptographic Algorithm

2.6.1 Symmetric Key Algorithms

Symmetric key algorithms are mainly utilized in the area of cryptography. The symmetric key algorithms are employed in plain text makes the process of decryption difficult, when the key is not shared. This is because, the same key is utilized for both the encryption and decryption and all the processes involve the same key. As the key is shared with the receiver, the confidentiality of the key is very important to ensure security. When the key is disclosed with other entities, then the security becomes a big question mark. The concept of symmetric key algorithm is depicted in Figure 2.4.

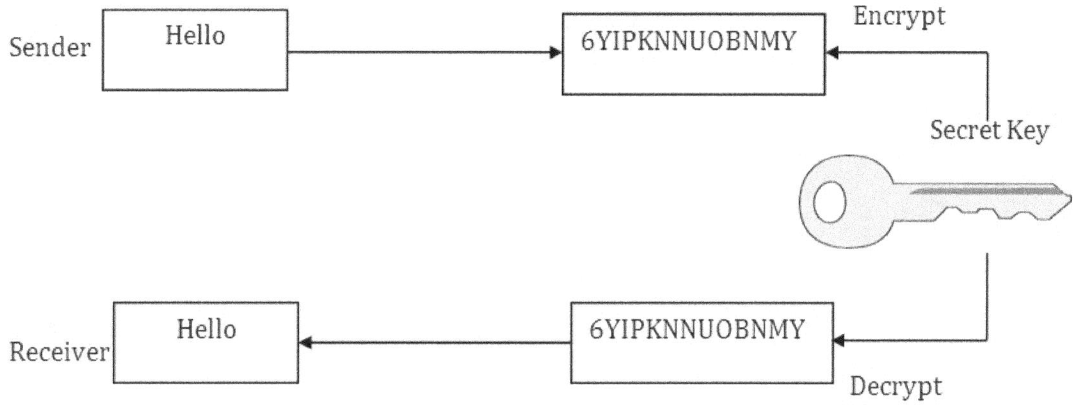

Figure 2.4: Process of Symmetric Key Algorithm

Symmetric key algorithms are usually incorporated to provide confidentiality or to produce Message Authentication Code (MAC). The symmetric key algorithms ensure confidentiality by utilizing the same key to attain both encryption and decryption. The MAC is produced by the symmetric key algorithms for ensuring authentication and integrity. The MAC can be produced by utilizing either the cryptographic hash function or the symmetric key algorithms. Some of the popular symmetric algorithms are DES, BLOWFISH, RC5, triple DES, AES and so on.

a DES

DES refers to Data Encryption Standard and it was developed in 1977. It is the first and foremost encryption standard being recommended by the NIST. DES involves 64 bit key with 64 bits block size. The DES algorithm accepts the plain text and converts it to cipher text without altering the length of the text. The block size employed by this algorithm is 64 bits. As already mentioned, the size of key is 64 bits, out of which 56 bits are utilized by the algorithm and the remaining eight bits are meant for parity checking. The general structure of DES is shown in Figure 2.5.

The process of decryption in DES involves similar series of operations as in encryption but the keys are employed in the reverse order. This algorithm involves sixteen rounds with Initial and Final Permutation (IP, FP). Initially, the block is decomposed into two with 32 bits by Feistel structure, which implies same procedure for both encryption and decryption, while the decryption is simply the reverse operation of the encryption with respect to subkeys.

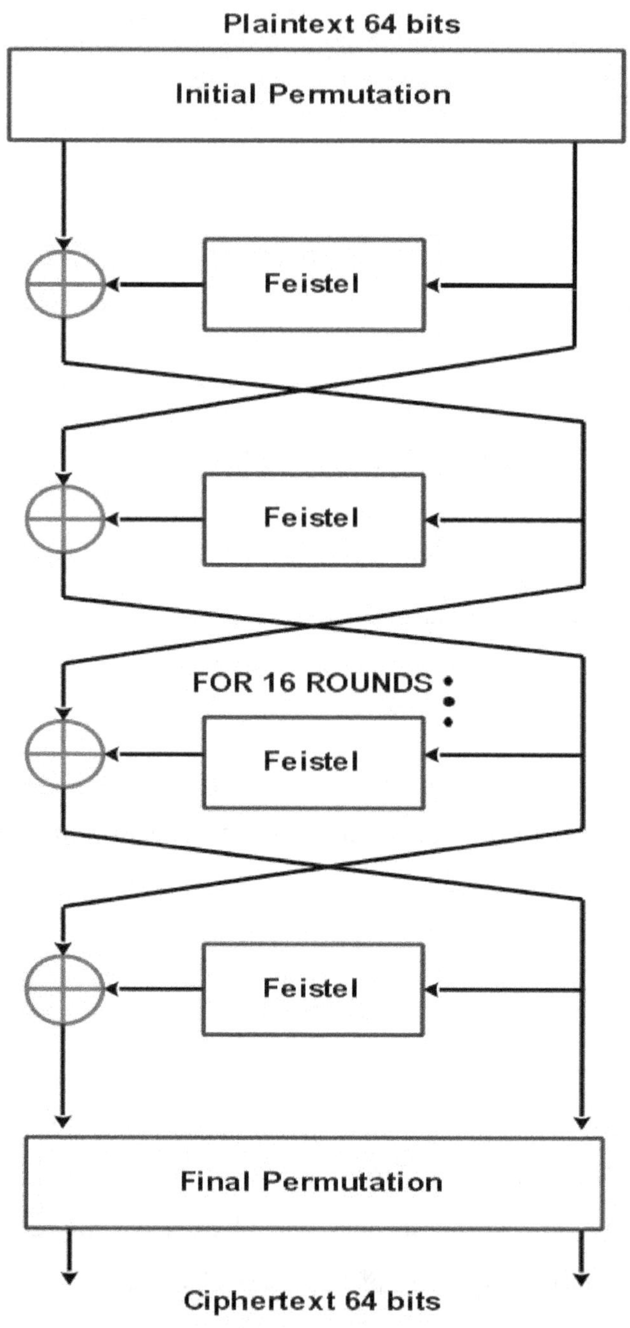

Figure 2.5: General Structure of DES

The Feistel structure involves four important phases such as expansion, key mixing, substitution and permutation. In the expansion phase, the 32 bit block is expanded by expansion permutation with 48 bits and the outcome is clubbed with the subkey by XOR operation in the key mixing phase. The block is then decomposed into eight 6 bits and passed to the Substitution boxes (S box). The S box substitutes the 6 incoming bits with four bits as outcome, which is a non-linear operation. At last, all the 32 outcomes of the S boxes are reorganized on the basis of fixed permutation by the P box and this process continues in every round. Hence, the algorithm is simple but effective.

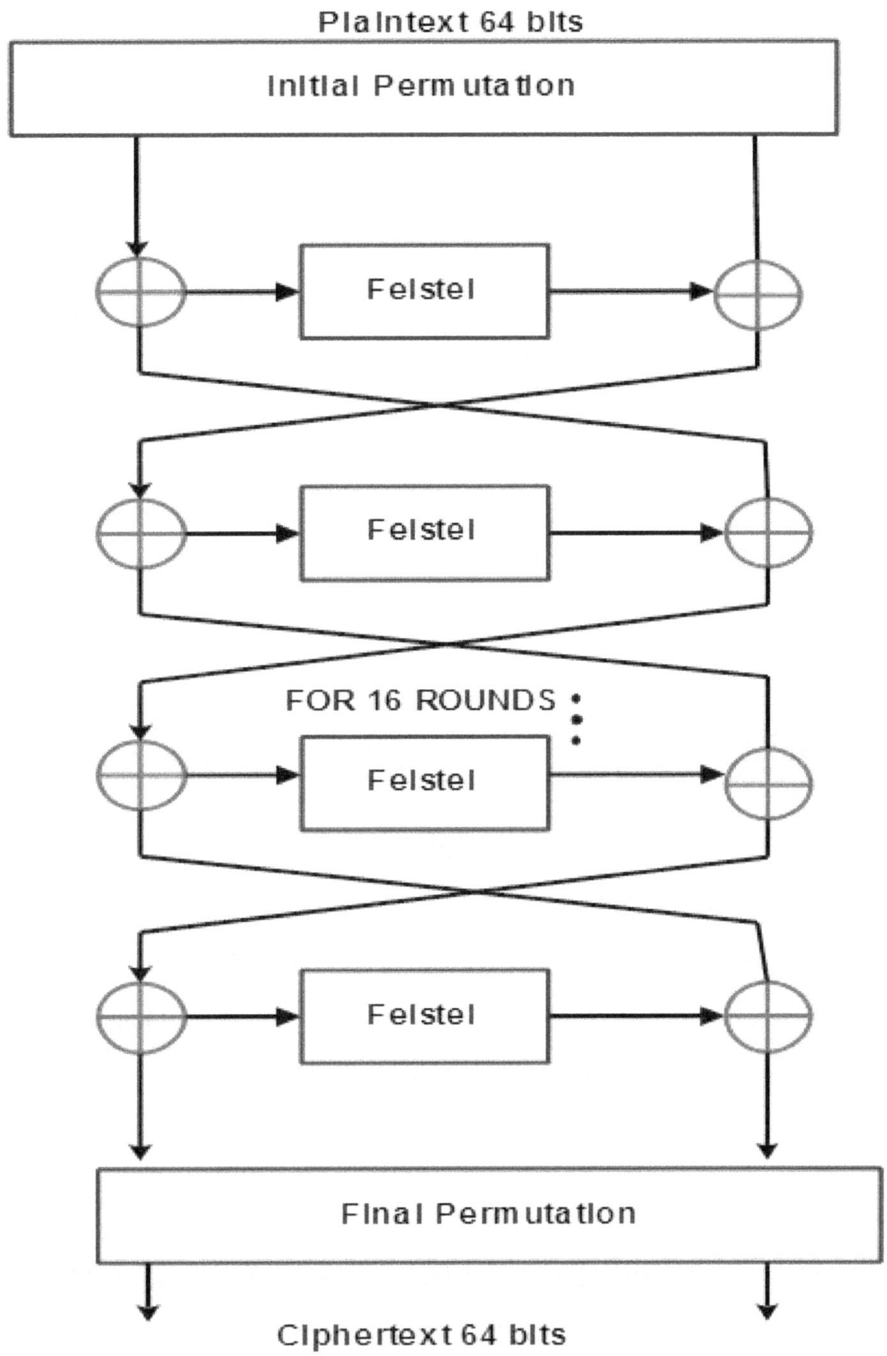

Figure 2.6: General Structure of Blowfish

b Blowfish

Blowfish is proposed in the year 1993. It is one of the famous encryption algorithms, which involves a variable length key and 64-bit block cipher. The length of key may vary between

32 and 448 bits. This algorithm involves 16 rounds of Feistel cipher with numerous s-boxes. The general structure of blowfish algorithm is shown in Figure 2.6.

Every round performs four steps for ensuring better security. Initially, the input is divided into eight bits and these bits are passed to the S-boxes, which obtains 8 bits and generates 32 bits. The outcome of this step is performed modulo and XOR operations to generate the 32 bit. This process repeats and in the 16th round, swapping operation is carried out. The decryption process is simply the reverse of the encryption process. Blowfish algorithm is so popular owing to its minimal processing time, throughput and efficiency.

c RC5

RC5 is developed in the year 1994. The key length of RC5 is MAX 2040 bit with a block size of 32, 64 or 128. Though this algorithm offers better security, the performance of the algorithm consumes more time for execution. RC5 is based on feistel structure but the key handling is difficult. The encryption and decryption processes of RC5 expand the random key with 2(r+1) words.

d Triple DES

Triple DES is proposed in the year of 1998, which is an improvised version of standard DES. In this case, the encryption method is the same as the original DES, however the algorithm is applied thrice increase the strength of encryption. However, this algorithm is a slow performer when compared to the other block cipher methods. It handles the 64 bit block size with 192 bits key size. It consumes more power and time when compared with DES.

2.7 AES

AES is a benchmark encryption standard being recommended by NIST as a replacement of DES. Brute force attack is the only proven attack against this algorithm. In the brute force attack, the attackers try different combinations of the character to break the security. The AES algorithm is depicted in Figure 2.7.

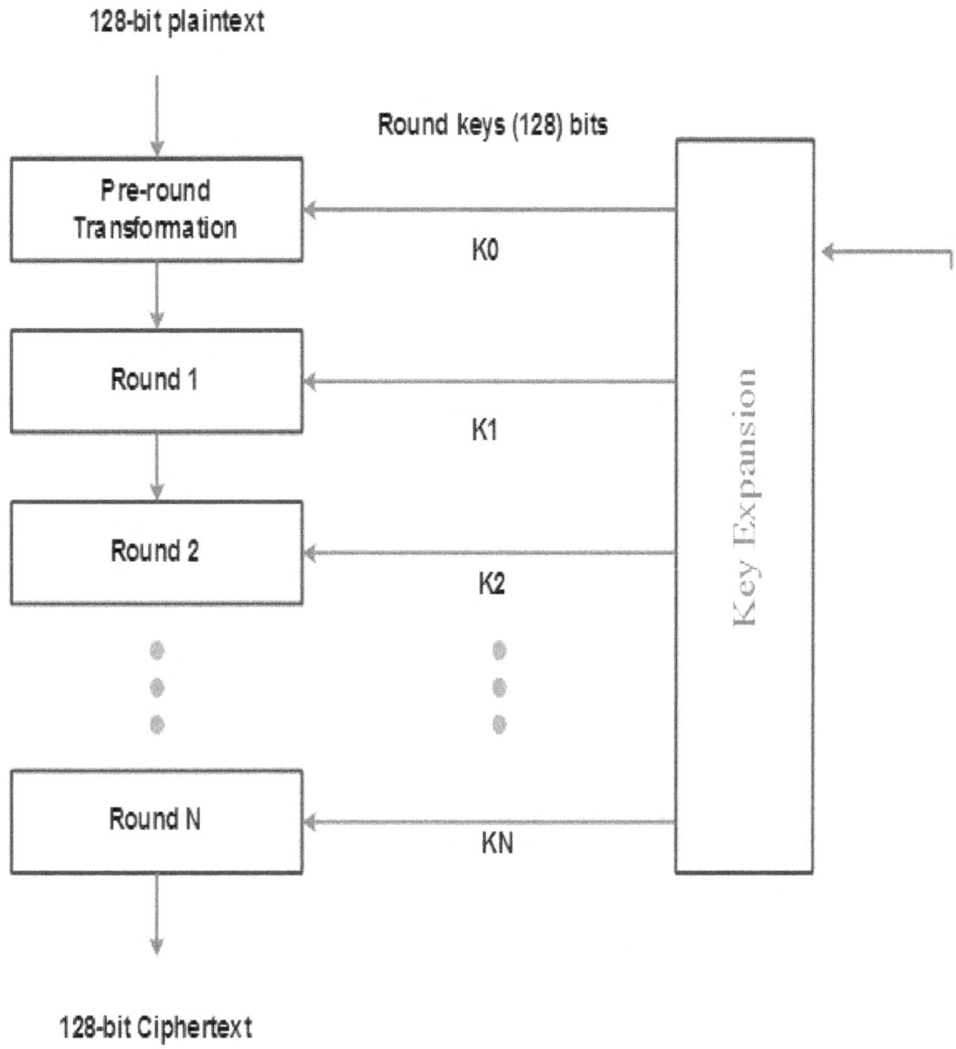

Figure 2.7: AES Algorithm Model

The variable key length of the algorithm can be of 128, 192, or 256 bits; default 256. It encrypts data blocks of 128 bits in 10, 12 and 14 rounds depending on the key size. In order to specify the key size, the AES algorithm is represented by AES-X, where X is the key size. Some of the noteworthy characteristics of AES algorithm are listed as follows.

The AES algorithm is flexible and combats against cryptanalysis better. This algorithm is compatible with contemporary processors and is applied even in smart cards. The AES algorithm is based on the Galois field theory and it relies on permutation and substitution techniques. AES encryption is fast and flexible and hence it is suitable for any application.

Hence, the outline of symmetric key encryption algorithms is presented and the following section highlights the asymmetric key encryption algorithms.

2.7.1 Asymmetric Key Algorithm

The asymmetric algorithm is based on a pair of keys called public and private key. The asymmetric key algorithm is also called as public key algorithm, as it involves public and private keys. The public and private keys are meant for performing encryption and decryption. In this kind of algorithm, the public key is shared with the users, while the private key is kept secret. The concept of asymmetric algorithm is depicted in Figure 2.8.

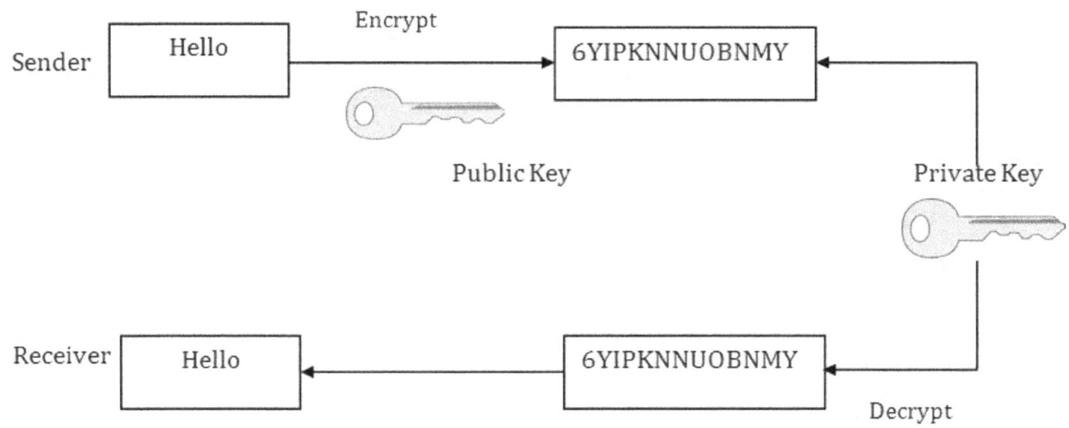

Figure 2.8: Principle of Asymmetric Algorithm

The association between the keys must be minimal and the utilization of public key must not support the detection of the secret key. Usually, the public and private keys are formed by the trustworthy agents of the key owner. The popular asymmetric algorithms are RSA, DSA, Diffie-Hellman and so on. A short note on symmetric cryptographic algorithms is presented as follows. Assymmetric algorithms are usually employed to form digital signature or to generate random numbers.

a RSA

RSA is utilized in Internet encryption and authentication and is developed in 1977 by Ron Rivest, Adi Shamir, and Leonard Adleman. The RSA algorithm is one of the most popularly utilized encryption algorithms. This algorithm is employed for both private and public key generation and encryption. RSA algorithm is proven to be faster.

The RSA algorithm is based on two prime numbers such as P and Q and the N is computed by multiplying P and Q. The public key is chosen such that it cannot be acquired by (P-1)×(Q-1), while the private key is chosen in such a way that the condition (D×E) mod (P-1)×(Q-1) is 1. The ciphertext is computed by $PT^E \ mod\ N$, which can be decrypted by

CT^D *mod N*.

b Elliptic Curve Cryptography (ECC)

ECC is an asymmetric algorithm developed by V.Miller (IBM) and N.Koblitz (University of Washington) in 1985. This algorithm makes use of different keys to encode and decode the data. ECC is a mathematical approach which provides a strong security with 164-bit key as other systems achieve with a 1024-bit key. The main advantage of ECC is that its utilization of small key lengths which results in quick encoding and consuming minimal energy.

c McEliece

In 1978 Robert McEliece proposed a public key cryptosystem called McEliece. It is based on the theory of algebraic codes and was not adopted by the implementers mainly due to large public key sizes. This algorithm uses the existence of a certain class of error- correcting codes called Goppa Codes. The idea is to hide the structure of the code by means of a transformation of the generator matrix.The transformed generator matrix becomes the public key.

Hence, this section outlines both the categories of security algorithms such as symmetric and asymmetric encryption algorithms. The following section studies and discusses the related works with respect to security in cloud data.

2.8 RELATED WORKS

This section contemplates to review the related literature in different perspectives of secure cloud data storage.

2.8.1 Data Security in Cloud Storage

The progress in cryptographic hash functioning techniques flourishes the techniques to break the hash functions. SHA is a set of hash functions approved by NIST and the initial algorithm SHA-1 used 160 bits.

As cited by Aumasson and Bernstein [2012], the familiar SHA-1 algorithm has been broken by Wang . Thus, the NIST planned to check the current status of cryptographic hash algorithms. This led to the introduction of a new family of hashing algorithms, which is SHA-2.

SHA-2 algorithms are designed to work with four different block sizes such as 224, 256, 384 and 512. When compared with SHA-1 group of algorithms, SHA-2 algorithms are more secure and are appropriate for commercial purposes. The major drawback being cited by the SHA-2 algorithms is the interoperability, as claimed by refGuo *et al.,* [2011].

The SHA-2 algorithms started to face several attacks and thus, the NIST again circulates a proposal for finding new secure and compatible algorithms. Several algorithms were proposed as members of SHA-3 and they are BlAKE, SKEIN, Keccak, JH and Grostl. BLAKE algorithm entered the final round of SHA-3 algorithm selection and is a family of four different hash functions such as BLAKE-224, BLAKE-256, BLAKE-384 and BLAKE-512.

As claimed by Aumasson and Bernstein [2012], Guo *et al.,* [2011] BLAKE makes use of minimum amount of resources and is faster both in terms of software and hardware. However, the security of this algorithm is not up to the mark.

As cited by Guo *et al.,* [2011] the Groestl algorithm is entirely different from the general design principle of SHA group of algorithms and is based on bytes. Some of the noticeable features of Groestl algorithm are its security, simplicity, side channel resistivity and parallelism. However, this algorithm is susceptible to collision attacks. The detailed description of the Groestl algorithm can be found in the work of Guo *et al.,* [2011].

The other important member of SHA-3 algorithms is JH algorithm and it consists of four different hash functions. They are JH-224,256,384 and 512. The major operations involved in JH are permutation, substitution and XOR. The design of JH is elaborated by Guo *et al.,* [2011] and the performance of JH is poorer than keccak. SKEIN which is a constituent of SHA-3 family algorithms comes in three different sizes, which are 256, 512 and 1024 bits. The building blocks of SKEIN algorithm are Three fish, Unique Block Iteration (UBI) and Optional Argument System, as cited by Guo *et al.,* [2011].

Keccak algorithm is the most efficient algorithm in the SHA-3 group of hash algorithms. This algorithm relies on the process of constructing sponge. Keccak is proved to be secure against generic attacks and some of the key merits of keccak are simplicity and flexibility. The keccak algorithm is well explained by Guo *et al.,* [2011].

Buyya *et al.,* [2009] present the perception, the knowns and unknowns of the services provided by cloud computing. The authors of this paper stated that cloud computing technology will soon become an essential component for leading life. Business based resource allocation frameworks are presented for cloud formation and is achieved with the help of Virtual Machines (VMs). The business based resource management schemes are proposed by focussing on the client-based service management and risk management with respect to the SLA based resource allocation. Additionally, the Industrial representative platforms are also discussed.

Wang *et al.,* [2011] discussed about the serious security challenges being faced by the commercial public cloud and analyses the security solutions of a trustworthy public cloud environment. This article discusses about the computational ability and cost effectiveness of cloud computing platform, in addition to the process of handling malicious cloud users from misusing cloud resources. The misuse can be anyone of the password or key cracking, malicious data hosting, or botnet command and control.

Buyya *et al.,* [2008] presented a business oriented cloud computing for providing service utilities. A wider perception about the formation of global cloud exchange for service trading is discussed. The recent cloud technologies provide restricted support for business oriented resource management and thus certain beneficial facilities such as negotiation mechanisms, allocation algorithms and risk management can also be provided. The interaction protocols should also be enhanced to provide interoperability between multiple CSPs. Internetworking concept is yet another issue in cloud, which could form a business oriented global cloud exchange for service trading. Additionally, the regulatory and legal issues faced by cloud computing are also discussed.

Bernstein and Vij [2010] have focussed on the security considerations in intercloud and reviewed the cloud computing association. Different facets of security in intercloud federated environment are considered in this article and a trust model is presented in association with the common Public Key Infrastructure (PKI). Additionally, the management issues with respect to intercloud environment are also discussed. This article claims that the management activities must be refined to enjoy better benefits.

Takabi *et al.,* [2010] present the challenges with respect to security and privacy in the cloud computing platform. The cloud computing platform with respect to different domains can utilize different security, privacy, trust requirements and employ various mechanisms, interfaces, and semantics. These domains could enable the services or other infrastructural or application components. Service Oriented Architecture (SOA) is the optimal technology to enable multiple domain formation by means of service composition and orchestration. This article concludes that the existing security and privacy remedies are needed to be reanalysed for their suitability to the cloud computing environment.

Dillon *et al.,* [2010*b*] discusses about the issues and challenges faced by cloud computing.

The important cloud computing paradigms such as service-oriented and grid computing are discussed in addition to the relationships between them. These challenges are considered with respect to the cloud computing. The authors of the article claimed that the cloud computing defines the way through which the users and applications interact through the cloud. This idea rules out the capability of users to pick other cloud vendors for resource optimization in various degrees. However, it is highly complex to club the cloud services with the existing systems of the enterprise.

Mei *et al.,* [2008] presents a tale of clouds with some paradigm comparisons and recent research issues with respect to cloud computing combined with service and pervasive computing. This article lists out the following contributions.

- The input/output characteristics of cloud computing are similar to the characteristics of service computing.
- The storage characteristic or principle of cloud computing is closely related to the pervasive computing than the service computing.
- The computation features of all the three paradigms are similar to each other.

Pearson and Benameur [2010] presented a paper that discusses about the privacy, security and trust issues of cloud computing and considers the security, trust and privacy issues of cloud computing and the means to fix these issues are discussed. The CSPs must ensure the privacy and security of the cloud data in place of the organization and the users. Specifically, it is important to assure that the privacy and security is not compromised by any means.

Hence to achieve this, multiple layers are involved and they are policy, process, legal and technological. This is the best idea through which the security is ensured throughout the life time of the system. When concluding the article, the authors state that the research with respect to private data security must be enhanced by considering the identity management and data obfuscation.

Wang *et al.,* [2008] presented an article that focuses on cloud computing with its definition and experience by including the details of dynamic and customized computing environments to the cloud service users. The recent advances of cloud computing with the related ideas and features are discussed. At last, a sample cloud for data centers is illustrated.

Cunsolo *et al.,* [2009] presented a new computing paradigm, which provides a solution for forming clouds out of heterogenous resources. This work provides a mean to the academicians or users to share the available resources and the computing resources can be sold as well.

However, this work suffers from major challenges such as reliability, security, interoperability, compatibility, QoS ad so on.

Zhang *et al.,* [2013] presented a work based on cloud security meant for both cloud users and providers. Different challenges of work such as IDM, data security, and virtualization security are examined. The cloud service users and the providers possess different security requirements and hence, the security solutions must be formed by considering this point. The disadvantages of the cloud computing can be overcome by the following suggested ideas.

- The service providers have to consider the safety of the cloud platform when compared to the classical computing platform, while maintaining the service continuity.
- The service providers must provide data security even during storage, transmission and recovery.
- The service providers have to offer the cloud service users with a better vitualization policy with security provisioning.
- The service providers must be vigil against the occurrence of security attacks and should provide a way to handle the attack without any loss.
- The service providers have to detect the application security issues with respect to varying service models such as IaaS, PaaS and SaaS, as discussed in the previous chapter.
- The service providers have to take the legal problems and the cloud users' comfort must be considered while implementing any new strategy.

Finally, the authors state that it is quite difficult to attain data security and data privacy of cloud computing data.

Viega [2009] presented a paper that relates the cloud computing environment with the common man. This work completely focuses on the user side which utilizes the resources via network. While using such resources, the users tend to face numerous doubts and questions about the data security and privacy, as the data leakage is a serious threat. When a cloud has undergone security breach, then the users of the cloud must be safeguarded to the maximum. Though the cloud computing paradigm supports the users in terms of cost effectiveness, efficiency, computational efficiency and so on, the cloud service providers are confronted by numerous risks or challenges such as ensuring data availability, data security, privacy and so on. Hence, all security requirements must be standardized.

Jensen *et al.,* [2009] discuss the technical security issues happens out of the cloud service utilization with the technologies meant for building the networked collaborations.

RefKretzschmar *et al.,* [2011] examined the different cases and situations of intercloud both in the public and private sector such as Distributed Management Task Force (DMTF), NIST, Global Inter-Cloud Technology Forum (GICTF) and European Network and Information Security Agency (ENISA). This work involves analysis and summary of requirement of security management with the help of identity management, credential management, attribute management, privilege management, digital policy management, configuration management, cryptographic key management, metadata management and audit management. All the aforementioned points must be considered for future enhancement of cloud security.

Zhou *et al.,* [2010] presented a review article focussing on the security and privacy of cloud computing environment. This work states that the main reason for the limited usage of cloud computing is the security and privacy issues though they are taken care of by the CSPs. The security and privacy solutions for cloud computing must provide certain other performance metrics such as availability, confidentiality, data integrity, control and audit. The law of privacy must be modified such that the bond between the cloud service users and providers is made better.

Harauz *et al.,* [2009] presented a model for securing the data in the cloud computing environment. Different measures such as data confidentiality, integrity, and availability must be considered by the CSP and must be fixed with the help of data encryption scheme. Besides, the access control must also be ensured to avoid malicious access to the data along with secured storage. Hence, the term security is the sole responsibility of the CSP along with other issues such as e-discovery, regulatory compliance, privacy, auditing and so on. All these issues are addressed by a security model with CIA.

Zhang and Zhou [2009] presented an Cloud Computing Open Architecture (CCOA), which relies on seven important principles and ten architectural modules, by combining the ability of SOA and the concept of virtualization with respect to both hardware and software. The authors claimed that the proposed CCOA provides a better configurable architecture for supporting nominal guidance, while presenting the infrastructure, software, application and the business process in a combined form.

Subashini and Kavitha [2011] have discussed several security related issues in different service delivery models of cloud computing. This article stresses more on security concerning the data privacy and protection. The authors suggest the users to be cautious against the security breaches involved in cloud computing. Finally, it is concluded that the cloud services are not only meant for internetworking services but for technological environment too. Numerous issues are still being addressed by the cloud computing environment in terms of SLAs, security, privacy and energy efficiency. Hence, the absence of an efficient security model hinders the real applicability of cloud computing and the level of security depends on the sensitiveness of the data. Besides, the cloud computing environment is based on heterogenous systems and presenting a single security system could not meet the goal for sensitive applications. Thus, the security system must be customizable such the application itself can alter the security policy.

AlZain *et al.*, [2012] presented a survey of security issues in cloud, as the cloud users are concerned about the sensitive information, which needs to get protected from the malicious users of the cloud. Additionally, availability is another main issue faced by the cloud computing environment. The authors of this work suggest the users to utilize multi-clouds for minimizing the security risks involved in the cloud computing environment.

Kim *et al.*, [2010] present certain security and architectural issues of cloud computing environment, which stresses on the national security community clouds with varying security features unlike the normal community cloud for providing support to Community-of-Interest (COI) requirements. This work deals with the needs and features of national security community clouds and the needs of COIs as well. The authors state that multi-cloud hybrid architecture meant for the national security cloud could store sensitive data in private clouds and insensitive data in the public clouds. Data sharing and collaboration is achieved with the help of community clouds. The community clouds can be divided into two, which are logical and physical instances with different functionality.

Jing and Jian-Jun [2010] present a security model for cloud computing after analysing that the standard security policies are not suitable for the cloud computing platform. The authors discuss about the security related issues and avoiding the security attacks. Finally, this work concludes that the reliable security solution is the current need of cloud computing now-a-days.

Kandukuri *et al.*, [2009] presents different cloud security problems and the cloud vendors are suggested to offer a strong assurance for SLAs in order to encourage the cloud users against security issues. However, it is really hard to promise security in the computing environment like cloud, while it offers different range of services. Each service model is susceptible to different kinds of security attacks and the SLAs must include all the different kinds of services along with the security policies. The SLAs must be standardized such that it is easy for both the provider and the user parties in offering and utilizing the cloud services.

Moreno-Vozmediano *et al.,* [2012] have reviewed different challenges involved in cloud computing. The authors state that the cloud computing stays in a considerable position in the future networking concept, enabling on-demand provisioning of applications, platforms, and computing infrastructures. Finally, the authors concluded that the cloud environment has to meet multiple technical challenges such as dynamic service provisioning, QoS, SLA negotiation, scalability, management, billing and payment; and context-aware services, interoperability, portability, availability, reliability, sustainability and energy efficiency.

Garber [2012] discusses about the data security, privacy, policy and dependability in cloud computing environment. The author highlights the concerns and the profit or loss of the service provider. The sample scenario is explained by Oregon Hay Products, which is a community bank for U$250,000 and some amount is stolen by cyber theft in the year 2010. The count of cyber thefts is increasing every day and hence, the security solution must be able to handle all the security related issues.

Monfared and Jaatun [2011] present details about monitoring the data intrusions and security breaches in the distributed cloud environments. Cloud security is the major challenge faced by the cloud computing environment. This work analyses and compares different existing security managing mechanisms along with the drawbacks observed in the techniques. This work presents a cross-layer security monitoring solution, which could deal with different challenges in the cloud computing model. However, a wide range of obstacles are still present in the cloud environment and the CSPs must access the components of cloud environment such that effective solutions can be built up. The authors state that the cloud environment still suffers from stringent data management strategies. The open communities are suggested to present standard components for ensuring the security and the security requirements as well.

Meng and Liu [2012] present an enhanced Monitoring-as-a-Service (MaaS) for managing the cloud environment better. This work states that Monitoring-as-a-service is an essential functionality for better cloud management. The authors stated that MaaS has to support all the usual state monitoring abilities commonly utilized cloud data centers in addition to the performance related features for attaining minimized monitoring cost, better monitoring scalability, strengthening the effectiveness of monitoring service consolidation and isolation. The authors propose three improvized MaaS functionalities as pointed out below.

- The instantaneous violation detection is replaced by window-based violation detection, which performs better.

- The period state monitoring is presented in place of violation-likelihood based state monitoring.
- Single tenant approach is suggested a replacement by means of multi-tenant state monitoring techniques.

The authors concluded that different issues are needed to be addressed with respect to cloud MaaS for attaining better cloud computing that ensures robustness, fault tolerance, security, privacy and elasticity to meet different and the changing needs of state tracking and state-based resource management with respect to a variety of cloud applications and cloud service provider. The follwing section presents the related works concerning the data integrity preservation for cloud data.

2.8.2 Data Integrity Preservation for Cloud Data

Data integrity is one of the major concerns of the cloud computing environment and so many research solutions are presented in the literature. Li *et al.,* [2017] checks the cloud data integrity by the identity based auditing scheme. The fuzzy identity based auditing protocol by employing the biometrics.

Yu *et al.,* [2016*a*] verifies the cloud data integrity by a TPA, which can perform audits over the outsourced cloud data. A remote data integrity checking protocol is proposed by Yu *et al.,* [2016*b*], which is based on identity based encryption. The key of this work is formed by homomorphic cryptography and is proven to be secure.

A public integrity checking scheme is presented by Zhang *et al.,* [2016*c*], which involves lightweight computation. Additionally, the proposed scheme works on batches of data and supports dynamic operations. A public verifying data integrity scheme is proposed by Zhang *et al.,* [2016*b*], which utilizes random data masking technique to combat against intruders.

Wang *et al.,* [2016*a*] presents a remote data integrity checking scheme based on identity based public key cryptography. An identity based data outsourcing scheme is proposed by Wang *et al.,* [2016*b*], which performs comprehensive auditing. The auditing is performed over the type and consistency of data.

An auditing framework is presented by Feng *et al.,* [2016], which is proven to be secure. Additionally, the proposed auditing protocol provides bidirectional authentication, error response system and follows a better load balancing strategy.

Le *et al.,* [2016] present a cryptography based remote data integrity checking scheme.

This work combines homomorphic message authentication code and a secure encryption scheme to preserve the integrity of the cloud data.

A cloud data integrity verification scheme is presented by Thosar and Sonewar [2016], which is based on BLS signature. A TPA is employed for checking the data integrity and it is achieved by batch auditing. A data integrity verification system is proposed by Haoyu *et al.,* [2017], which is based on multi-branching tree. This work claims that it supports dynamic operations over cloud data.

A data privacy and integrity checking scheme for spatial data is presented by Ba-Hutair and Kamel [2016]. This work introduces a new index structure that is based on z-curve. Zhang *et al.,* [2016*a*] present a dynamic hash table based public auditing protocol. The performance of this protocol is checked by introducing a malicious cloud server, which deletes or modifies the cloud data. However, the protocol can address the issue perfectly.

A probabilistic verification technique is proposed by Rong *et al.,* [2016], which is termed as Verifiable Secure kNN. The proposed technique exploits the algebraic properties of the scalar products in encryption and correctness checking. Li [2017] proposes a novel availability and integrity verification protocol. This protocol ensures cloud data availability and integrity by means of fractional dynamic data sustenance.

2.8.3 Replica Management in Cloud Computing

Li *et al.,* [2015] propose a technique to minimize replication based on proactive replica checking. The objective of this work is to reduce the cloud storage space by creating minimal replicas. This work claims that the reliability is maintained with minimal data replication. However, this work does not follow a concrete idea for minimizing the rate of data replication.

A data replica placement strategy based on genetic algorithm for cloud is proposed by Cui *et al.,* [2015]. This work utilizes a tripartite graph model to construct the data replication issue and is tackled by genetic algorithm. However, this work involves computational complexity.

A replica placement algorithm is proposed for decentralized edge computing environment by Aral and Ovatman [2018]. This work is based on dynamic creation, replacement and deletion of replicas by considering the data requests. The shortcoming of this work is that the concept of dynamism introduces several computational overheads. A QoS aware data replication technique is presented for data intensive applications for the cloud computing system by Lin *et al.,* [2013]. This approach presents two different algorithms for dealing with data replication related issues. Initially, the greedy algorithm is applied but the replication cost cannot be reduced. Hence, the data replication issue is converted to Minimum-Cost Maximum-Flow (MCMF) problem. However, this approach suffers from time complexity.

Zhao *et al.,* [2018] propose an optimal replica distribution scheme for edge node assisted cloud peer to peer environment. The edge nodes are utilized for their increased upload bandwidth and these nodes are employed for handling the replicas. The edge nodes deal with the user request. However, this work consumes more resources of cloud.

The identity based public multi-replica provable data possession is proposed by Peng *et al.,* [2017]. This work presents a third-party verification for outsourced data that has multiple replicas in the absence of Public Key Infrastructure (PKI). However, this work does not concern about the data organization and any strategy for creating replicas.

Chen *et al.,* [2015] propose a work to reduce the joint response time for peer-assisted cloud storage systems. The varying type and size of data makes it tough to proceed with replicas being stored in various locations. This work considers a new metric namely joint response time that considers the delay in queue and service time. However, this work does not involve replica organization.

A new model to check for the consistency of the data namely Consistency as a Service (CaaS) is presented by Liu *et al.,* [2014*b*]. This work is based on two-levelled auditing architecture that needs a loosely synchronized clock of the audit cloud. The algorithms to verify the metrics such as commonality and read staleness are proposed.

Nahir *et al.,* [2015] propose a replication based load balancing technique. This approach creates multiple replicas for each and every job and the replica is stored in different servers. This work considers different queue length and load conditions. However, this work focus only on the delays and not the replica based operations.

A data replica placement mechanism meant for open heterogeneous storage system is proposed by Xu *et al.,* [2017]. This strategy takes the data availability, count of replicas and the load is balanced. However, this work does not consider the data based operations and the organization.

Hence, the related works with respect to data security in cloud storage, data integrity preservation and data replication management in cloud computing are explained and the following section concludes the chapter.

2.9 CHAPTER SUMMARY

This chapter summarizes the security issues being faced by the cloud computing environment and the possible ways to deal with the security issues. The security threats in the SaaS environment are also discussed. Additionally, the related works to three different issues of cloud computing are discussed.

CHAPTER 3

TSS - TWIN LAYERED SECURITY SCHEME FOR CLOUD DATA INTEGRITY PRESERVATION

Data storage is the most convenient service provided by the CSP, and the data owners could not find difficulty in issues related to managing and storing the data. Yet, the data owners are cautious about the data security. This chapter introduces a Twin Layered Security Scheme (TSS) meant for conserving the data integrity.

This chapter presents the introductory idea of the proposed TSS for cloud storage in section 3.1. The proposed security scheme is elaborated in section 3.2, which includes the hashing and encryption phases along with the sample work principle. Section 3.3 analyses the performance of the proposed work by varying the SHA3 member algorithms, Keccak versus Merkle hash tree and encryption algorithms.

3.1 INTRODUCTION

The technological world collects and utilizes huge volume of data for every moment of time. Out of the complete data, most part of the data is confidential and it may contain some sensitive information such as personal, health, financial information and so on. The data growth is observable with time and hence the data storage and organization become difficult for the organizations. Additionally, the organizations must upgrade the storage entities and they are expensive.

Managing the data is another inevitable exasperation for the organizations, though the principle of cloud computing helps the mid and small scale industries by presenting an economic friendly environment. Besides this, the cloud computing environment presents numerous smart features such as providing services on demand, options for payment of money in an easy manner, simpler data maintenance and so on.

All the above stated features inspire the data owners to contract out the business or personal data to the cloud storage, such that the data owners experience several advantages such as reduced overhead related to storage and data management, as cited by JoSEP *et al.,* [2010], Chen *et al.,* [2011], Wang *et al.,* [2014], Rahimi *et al.,* [2013]. In spite of these advantages, the data owners are still unwilling to contract out the confidential data to the

CSP, as the direct control of data may get missed, as cited by Wang *et al.,* [2010], Khan *et al.,* [2014*a*], Wang *et al.,* [2013], Wei *et al.,* [2010*a*] and Zhang *et al.,* [2014].

This leads to some security related problems such as confidentiality, integrity and availability. For example, some clouds may behave badly by eliminating or altering the occasionally utilized data, as cited by Ali *et al.,* [2015], Shamshirband *et al.,* [2013], Khan *et al.,* [2014*b*], Yang and Jia [2012].

A similar problem is observed in case of Amazon S3, Gmail and Sidekick for failure, email removal and disaster, as cited Armerding [2012]. Hence, it is mandatory to maintain the data consistency, while maintaining the integrity as well. Taking this as a challenge, this chapter attempts to present a security scheme that could conserve the integrity of the cloud stored data. The proposed security scheme preserves the data integrity of the cloud stored data with the help of Keccak's hash function and AES algorithm for performing the data encryption.

Hence, the proposed approach presents a TSS and the performance of the proposed scheme is verified by comparing the work with the SHA-3 member algorithms such as BLAKE, SKEIN and SHA2 in terms of throughput and memory. The time consumption for performing the encryption is also tested and the complete work is elaborated in the following section.

3.2 PROPOSED SECURITY SCHEME

This section presents the TSS for cloud data storage. The significant components being present in the proposed security model are Data Owners (DO), CSP and Integrity Checkers (IC). DO hands over the sensitive information to the cloud server and the CSPs are meant for offering a variety of services to the cloud users. This work focuses on the cloud data storage services.

In order to escape from the data management and storage issues, the DOs contract out the data to the CSP. Both the parties enter into a SLA, which defines the terms and conditions of the utilized service. However, there are chances that the SLAs may get broken at times and it introduces several issues. Hence, there must be an authority to check for the cloud stored data consistency and this task is carried out by IC.

In order to achieve data integrity, the complete work is divided into two parts namely hashing and encryption phases which are again divided into several sub phases. Initially, the

Keccak's hash function is employed over the data to be contracted out and the hash code is computed. The dynamic key is then produced, which strengthens the side of data owner, as this key is not revealed to anyone. The complete work flow of the proposed work is illustrated in Figure 3.1.

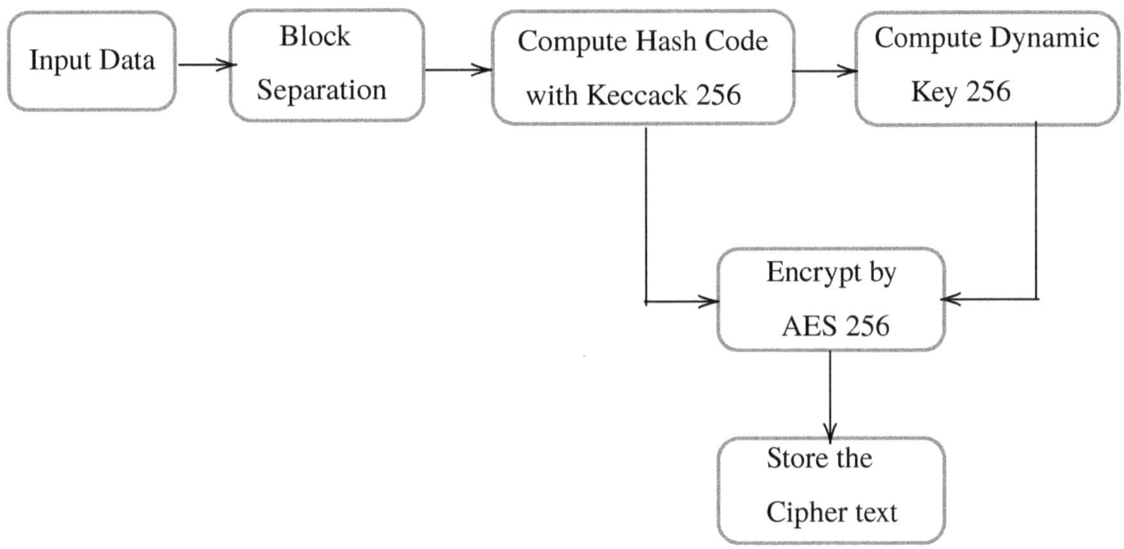

Figure 3.1: General Flow of the Work

The AES encryption algorithm processes by considering the so computed hash code and dynamic key and produces the encrypted version. Without the dynamic key data decryption cannot be achieved. Thus the security of the system is enhanced and the phases are explained as follows.

3.2.1 Hashing Phase

The main goal of hashing procedure is to improve the security of the system. Tree hashing is very advantageous, owing to the introduction of parallelism, quicker hash calculation and finally, the possibility to verify hash without having to check all the message blocks. The noteworthy tree hashing technique is Merkle hash tree. However, the crucial disadvantage of Merkle hash tree is the storage and retrieval from the tree constructed from the data owner's database.

In order to cope up with this issue, this work proposes to employ Keccak's hash tree. The important steps involved in Keccak's tree are sponge and duplex establishment. Sponge establishment intends to create a sponge function with required rate and capacity as parameters.

The goal of this function is to take the data as input, process it to obtain the outcome. The outcome of this function is the lumps of data in random sizes. This function takes the length of data in bytes. The reason for this is to avoid bit attacks.

While compartmentalizing the data in terms of bytes, there may be a chance to get remaining bits. The data which holds lesser than 8 bits is handled by a separate function. In order to initialize the sponge, two parameters are needed, which are rate (r) and capacity (c). It is a rule that the summation of r and c must be 1600; r+c=1600. However, it can be minimized with respect to the amount of data and computational speed. The proposed work fixes the value as r=144; c=256 (r+c=400).

In this step, the process works the same as sponge establishment, where the data bytes and the trail bits are separated. The isolated data bytes are assigned a pointer to the buffer and the trail bits are given a byte number. This is to avoid the manual copying of data blocks into buffer and then to append the data block with frame bits, which is followed by duplex call. In this work, the data blocks are just assigned with a pointer and the frame bits are passed in separately. This reduces the computational complexity.

The duplex function is initialized by the *r* parameter. The parameters of this function are (k(400), pad10*1, r). The value of r must be lesser than or equal to 400; $3 \leq r \leq 400$;

The purpose of hashing phase is to tighten the security of the system and is achieved by Keccak's hash function. This phase involves two phases such as block formation and calculation of hash codes. Keccak is a collection of hash functions and the main process involved in Keccak is the sponge establishment.

Bit rate (*br*), capacity (*c*) and diversifier (*d*) are considered as basic building blocks of sponge function and the sponge function relies on those parameters. The degree of permutation is determined by both the bit rate and capacity. The purpose of the diversifier is to distinguish between the Keccak entities. The main advantages of this work are it does not need table lookup and neither requires arithmetic operations nor data rotations.

The whole data to be contracted out is processed by Keccack's function by decomposing the data into several chunks by means of sponge function. Hence, the input data may incur any size and sponge function produces a group of data chunks, where the size of each chunk is chosen by the user. Hence, the bit rate and capacity together decides the system speed and security. The bit rate is directly proportional to the system security.

Conversely, when the capacity is increased the system speed is improvised. The proposed security scheme considers two output lengths such as 256 and 512 for analysing the performance of the system. Keccak imposes a condition that the sum of bit rate and capacity should be 1600. This Keccak function carries out all the functions in terms of bytes for dealing with the bit attacks. The hashing phase is pictorially presented in Figure 3.2.

This function functions on bytes and so it becomes problematic if it is possible to divide the input data into bytes perfectly. In such scenario, some count of bits is treated separately. The rest of bits are named as trailing bits and they were attached to the last node of the tree. Consider D as the data which had to be contracted out to the cloud. The proposed work organizes the data D into different data blocks or chunks and is represented by

$$D = DB_1, DB_2, DB_3, ...DB_n \tag{3.1}$$

These data blocks are structured as a standard binary tree. All the child nodes are processed for hash codes and were attached to the parent node. Repeat this process till the root node is reached.

$$H(D) = H(DB_1), H(DB_2), H(DB_3), ...H(DB_n) \tag{3.2}$$

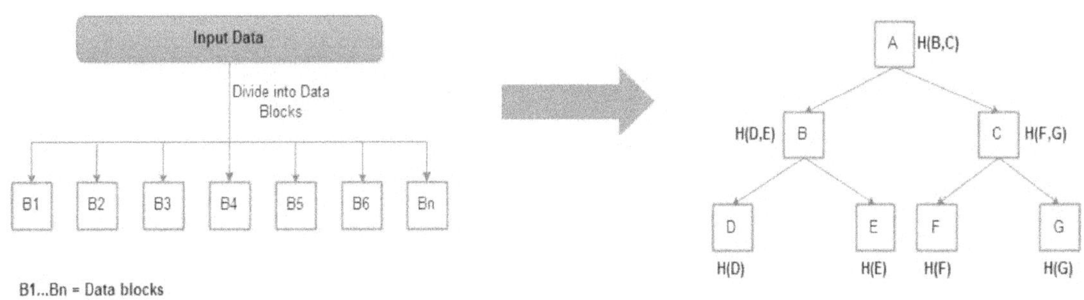

Figure 3.2: Hashing Phase

The complete hashing phase is presented as follows.

Algorithm 1 Hashing Phase

Input:data, ol

Output: Data chunks or blocks

Begin

Divide data with respect to ol;

If ((size of data)%8)==0) then

Construct tree;

Goto Hash code Computation;

Else

Collect trailing bits in a separate block;

End;

Algorithm 2 Hash code Computation

Input: Data chunks or blocks

Output: Hash code

Begin

Apply hash operation over leaf nodes;

Do

Merge the hash codes of nodes having same parent;

Repeat till the root node is reached;

End;

The input data is decomposed into data chunks, and a tree is constructed with multiple nodes. The tree grows by considering the size of the data and no restriction is imposed with respect to the height of the tree. As the size of input data increases, the height of the tree also increases. Thus they are directly proportional to each other.

In Algorithm 1, ol refers to the output length and trailing bits refers to the residual bits left after the decomposition of the input data into data chunks. At first, the hash codes are calculated for each and every leaf nodes. Clubbing of the hash codes are done when the leaf nodes are the offsprings of the same parent. Repetition of this process has been undergone till the root node is reached. In this way, the hash codes are calculated and are saved in the root node.

3.2.2 Encryption Phase

By applying the concept of encryption to the tree of hash codes, protection of data can be strengthened. So the next step is encrypting the hash codes to make it more secure. The encryption phase utilizes the most popular AES algorithm. AES is popular for its speed of execution and high standard security. With the help of a dynamic key, the hash codes are encrypted to enhance more security. The dynamic key is kept as secret and is not disclosed to anyone by the DO. The proposed work produces dynamic key of 256 bits by utilizing the CodeIgniter key generator.

The proposed work utilizes AES for encryption, as it works on bytes and not on bits. The proposed work employs a 256 bit key on AES algorithm and it involves 14 rounds of encryption. The greater the round count, the better is the security. Each round utilizes a 128 bit key being computed by the AES and the algorithm for encryption phase is presented as follows.

Every single round of AES is based on several important operations such as byte substitution, shift rows, mix columns and add round key. Firstly, the input bytes are passed into the S box in matrix format. The rows are then shifted by considering the positional constraints.

Algorithm 3 Encryption phase

```
1: Begin
2: For all data blocks or chunks
3: Do
4: Calculate dynamic key d by using CodeIgniter key generator;
5: Pass d and hash code to AES;
6: Return cipher text;
7: End;
8: End;
```

The columns are then jumbled up by substituting the original bytes into different bytes and this operation does not happen at the 14th round alone. The closing components of the matrix are done with XOR operation by a round key. This process is carried out for 14 rounds such that the cipher text is produced.

The dynamic key is not at all shared and hence, the security of the system is even better. Additionally, the plain text is not directly encrypted and in place of that the computed hash codes are encrypted. Hence, the twin layered security is provided by this work.

The DO can carry out integrity check by decrypting the AES algorithm with the help of dynamic key which is kept as secret for obtaining the hash code. The hash codes are then compared for checking the data integrity. The integrity check can be initiated by the data owner at any point of time by challenging the cloud server. The following section explains the proposed work with a sample.

3.2.3 Sample Work Principle of Twin Layered Security Scheme

To demonstrate the proposed concept, a little portion of data is extracted from the huge volume of data, as even a small portion of data has its role in a bigger tree of data. This part is included for better understandability of the concept.

The sample data being considered is of about 2886 bytes. For demonstrating the proposed work, the input data is divided into multiple 500 bytes. Hence, the tree is formed which contains 6 nodes and the last node contains the residual 386 bytes, as represented in Figure 3.3.

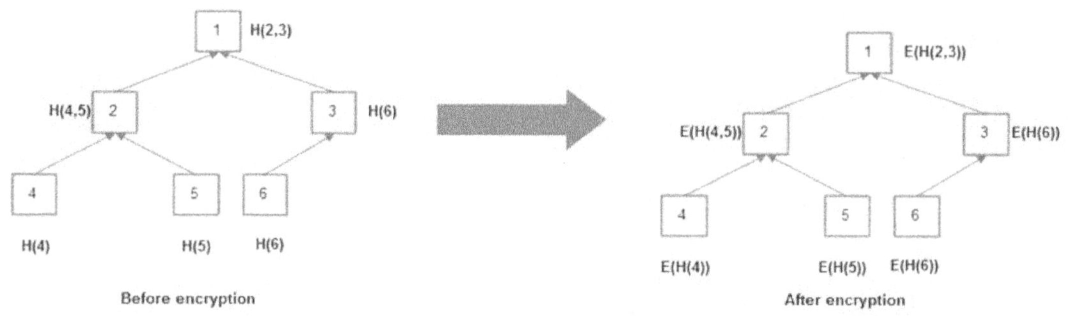

Figure 3.3: Structure of Sample Data

Initially hash codes are computed for leaf nodes. In the above figure, the nodes four, five and six are the leaf nodes. Calculate the hash code for sixth node and store it in the third node. Similarly calculate hash codes of fourth and fifth nodes and store it in the second node. The hash codes of second and third are stored in the first node, which is the root node. Hence, the whole tree is treated. The number of hash codes is minimized, as the hash code is calculated only for the child nodes with the same parent.

The root node saves the hash code of the entire tree and it doesn't contain any other data. Keccacks hash function of 256 bits is used to calculate the hash codes and then the dynamic key is generated by CodeIgniter and the hash codes are encrypted by the computed dynamic key with the help of AES algorithm.

Computed hash codes by Keccak (256 bit) [HC]

Sixth Node: bf807b6a14256c3e49c479060f83d24a941e4e9f3a1daadbc052e7e87c7079b7

Fourth and Fifth Nodes: 0f1b466a489726a7763cfb48c2443206556320ed174912a6429b0 8445058455a

Second and Third Nodes: bebd623270d374cf11e7dab1ca385822a793dcb2c 474ec0e98d0dd 2ee5cf23b1

Finally, the hash codes of the second and third nodes are stored in the root node. After this process, a dynamic key (256 bit) is produced as presented below.

Generated dynamic key [DK]: agC7Z5TKA6mi6trYCr7NeB4LE6mxqhug

AES encryption (256 bit) [HC+DK] :

Sixth Node: D/jDd8VKWvVByEcSaOrowYiEJMZpV5bJBp29IXYVWMqoySCxnwO GuEZ6zvSLiZttpvefR7Qvss4kkladqyKAUA==

Fourth and Fifth Nodes: hqq4Na122rFXfEB2+UE2sCvYexrcs0MosAedjrESVxPn6u/ Re8mcTcG4FNO/Vn9hMc6B5FMtMaZtIi+y+dPI7A==

Second and Third: 6gh7SOQc1wqn7iNmmrwfubvZmN7A9R95LAf/1Pk/BAt1SBJDm DWS8sr/G1Gfx5q5c7ubboiqEsuY7ZZLurDFPg==

The hash codes of the nodes can be computed only when the dynamic key is known. Even though the dynamic key is sniffed, it is tough to break the Keccak's security. No security breaches are reported for Keccak till now. Additionally, this work employs AES algorithm of 256 bits with 14 rounds for computing the cipher text.

Hence, the security of the system is proven to be better at the cost of complex computation. Several existing systems are based on the computation of hash codes alone. This work includes yet another line of security for better performance. The performance of the work is analysed and the results are discussed in the following section.

3.3 PERFORMANCE ANALYSIS

The outcome of the proposed work is examined in three different aspects. Firstly, the performance of the proposed approach is evaluated against the SHA3 member algorithms. The second round of performance analysis is done by comparing the proposed approach with Merkle hash tree. The third round of performance analysis is performed to justify the selection of AES algorithm.

3.3.1 Performance Analysis with respect to SHA3 Member Algorithms

The outcome of the TSS is evaluated against SHA 3 member algorithms such as BLAKE proposed by Aumasson and Bernstein [2012], SKEIN presented by Guo *et al.,* [2011] and SHA2 discussed by Guo *et al.,* [2011] with respect to throughput and memory consumption. Throughput determines the execution speed of the algorithm, which lays stress on the ability of the algorithm to deal with the data rate and is measured in terms of Mbps (mega bits per second). The effectiveness of the proposed system is also measured based on the memory consumption. Finally, the proposed work examines the time utilization for carrying out the encryption and the experimental results are furnished below.

In order to minimize the execution time, the speed of execution should be maintained greater. In case of minimal throughput, the execution process consumes more time, such that it consumes more resources as well. The proposed model is analysed for the execution speed by varying the input data and acceptable results are obtained.

The following graph represents the execution speed known as throughput for three different data with different sizes. From the graph it is clear that SHA 2 has very poor performance in all the cases. The performance of Keccack is good in all the cases and is followed by SKEIN. As Keccack processes the data in block wise, its throughput is higher when compared to others.

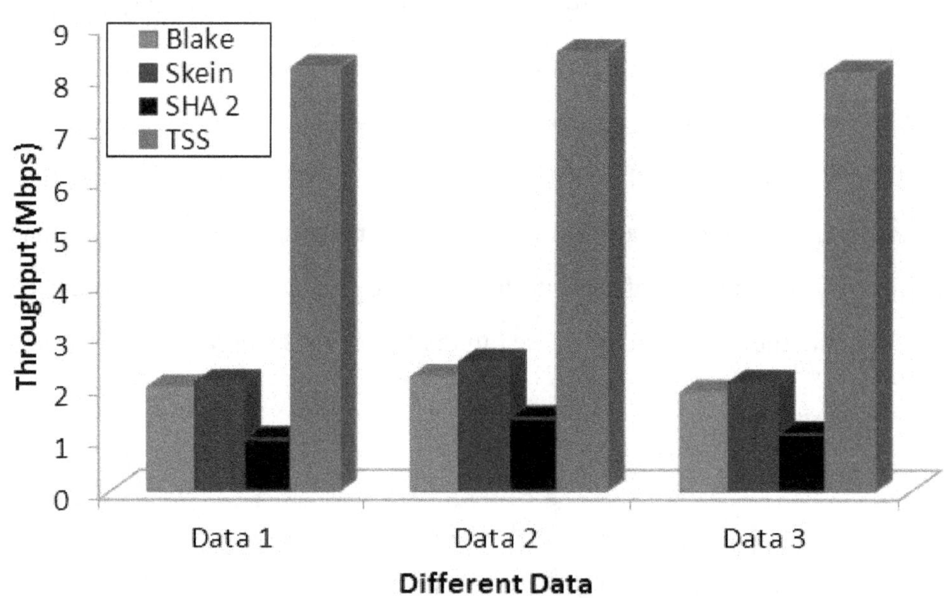

Figure 3.4: Throughput Analysis

Memory consumption is the other important performance metric to be examined. The proposed TSS analyses the consumption of memory by varying the different amount of data with the same comparative algorithms. The memory consumption of the proposed model is proven to be minimal. The consumption of memory is minimal due to the usage of data blocks and storing hash codes instead of original data.

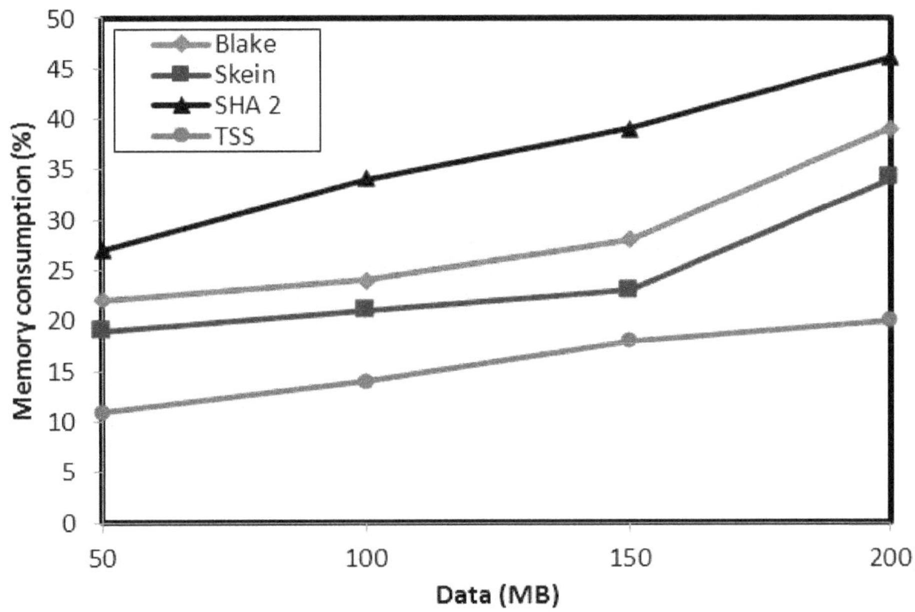

Figure 3.5: Memory Consumption Analysis

Besides this, the advantage of this proposed work is hash codes are calculated only for the children nodes of the same parent and is not needed to be calculated for individual nodes. All these features minimize the consumption of memory for the proposed work. The memory consumption analysis is presented in Figure 3.5.

The efficiency of all algorithms is also measured by an important performance metric known as execution time. The execution time should be as small as possible, in order to minimize the latency. A good performing algorithm having long execution time is infeasible. So it is good to minimize the execution time. The proposed security scheme's execution time is examined with different volumes of data. This experiment is performed with two different output length of 256 and 512 bytes and the results are analysed.

It is noted from the graph 3.6 that the length of output is a determining factor of the execution time. The above result is achieved with the Intel core i7 processor. When compared with 512 bytes, the output length of 256 bytes takes more time for execution.

If more importance is given to speed than security, then the output length can be increased. The br, c and d are set as 1088, 512 and 32 respectively for the output length 256 bytes.

Alternately, the values of br, c and d are set as 576, 1024 and 64 respectively for the output length 512 bytes. Hence, the security of the system gets enhanced when the count of rounds increases.

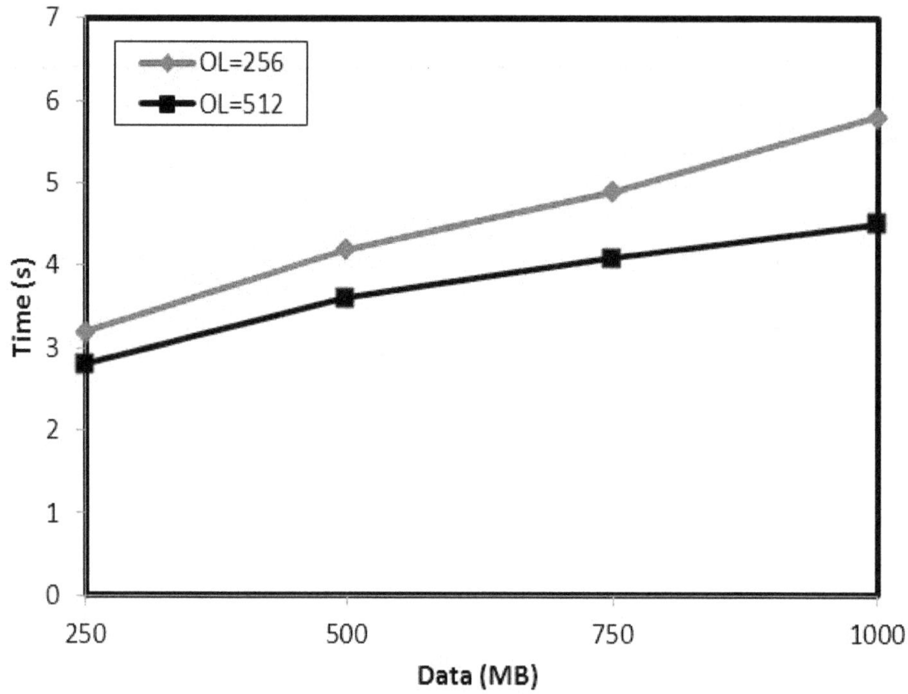

Figure 3.6: Execution Time Analysis

Hence, the main goal of the chapter is to propose a security scheme for attaining data integrity in cloud storage is achieved. The proposed security model is analysed by employing different performance metrics and the proposed work outperforms the comparative algorithms with better throughput, minimal memory and time consumption.

3.3.2 Performance Comparison between Keccak and Merkle Hash Tree

The existing auditing mechanisms suffer from several drawbacks. Few of them are communication overhead, authentication issues and so on. These issues are intended to be fixed by the proposed work with the help of Keccak's hash tree. The public auditing mechanism considers the degree of nodes, such that the verification process of multiple replicas becomes efficient and effective.

Keccak's hash tree processes every block of the data and it can process multiple blocks of data at a time. As parallel processing is made possible, the execution time is very much lower. The parallel computation can be classified on the basis of multiple or single message.

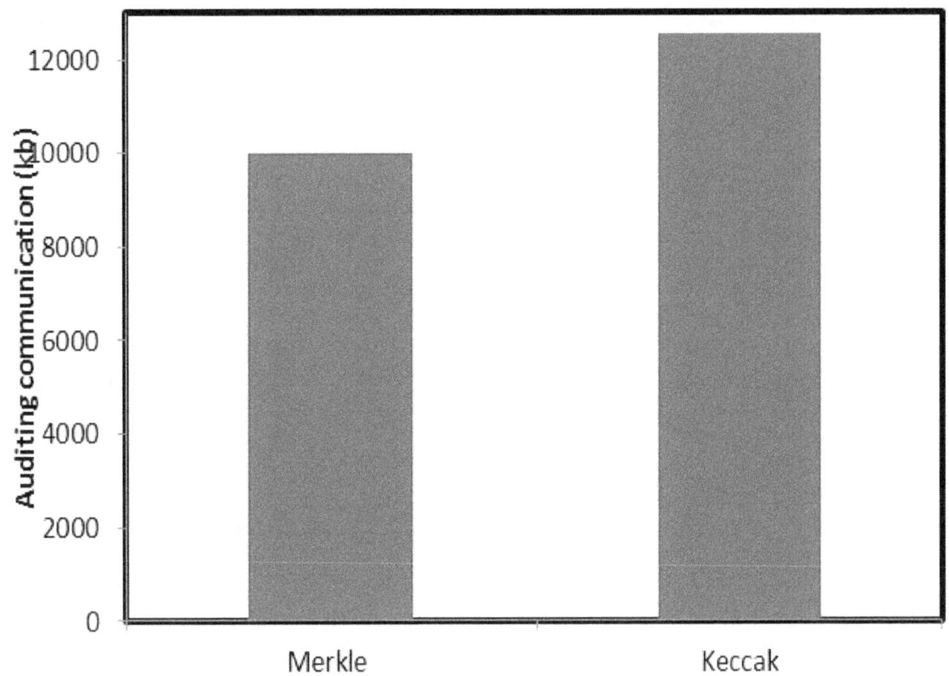

Figure 3.7: Auditing Communication Analysis

In the case of multiple message schemes, the data blocks which are sponges are not dependent over each other. If it is single message type, then the sponges are interdependent and hence the outcome of a sponge will be the input of another sponge.

The number of nodes in a Keccak's tree can be decided on considering the length of the data. Every node in the tree processes the sponge and the higher level node depends on the lower level node. The outcome is returned in the root node of the tree.

The inputs passed to this tree are height, degree, block size to be processed, rate and capacity. The parameter rate decides the speed at which the data has to be absorbed by the node. Capacity is the number of bits. Height is the number of levels that a tree can grow. Degree is the parameter that decides the number of children that a node can possess.

The growth mode of the Keccak's tree to be used in our work is leaf interleaving, as such tree can work for any sized input. In addition to this, the proposed work compares the performance of the Keccak is analysed against Merkle hash tree in terms of auditing communication, storage overhead and data transfer.

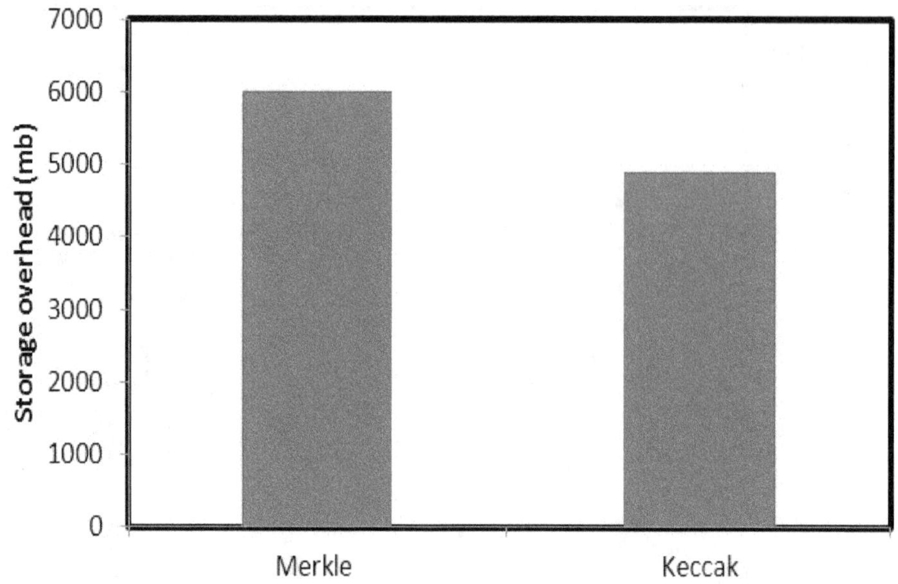

Figure 3.8: Storage Overhead Analysis

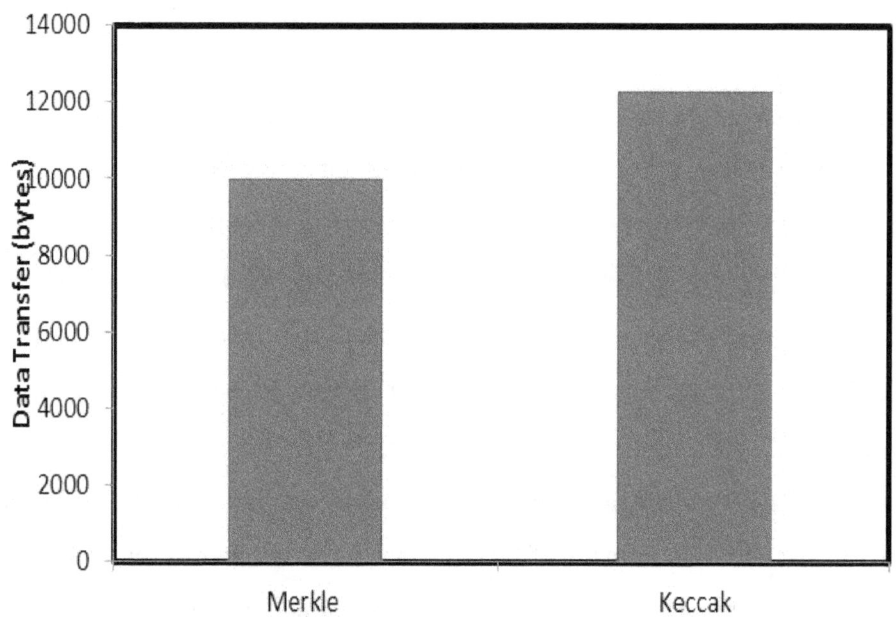

Figure 3.9: Data Transfer Rate Analysis

Auditing communication overhead is considered, as this work maintains a tree like structure with leaves. As the work has to consider all the data in the leaves, the auditing communication is a bit higher than the Merkle hash tree. On the other hand, the storage overhead is minimal for the proposed approach with better data transfer rate, as shown in Figures 3.7, 3.8 and 3.9. The proposed work is analysed to prove the efficacy of AES algorithm in the following section.

3.3.3 Performance Analysis by Varying Encryption Algorithms

This section aims to emphasize on the efficiency of the proposed security scheme in two ways. Initially, the choice of AES algorithm is proven by comparing with the analogous algorithms such as DES, Blowfish in terms of encryption time and throughput. Finally, the performance of keccak and AES algorithm is analysed by Avalanche effect.

In the proposed approach, AES-256 is utilized and the reason for the choice of AES-256 is justified by performing comparative analysis with the analogous encryption algorithms such as DES, Blowfish, AES-128. The efficiency of encryption can be judged with the help of Avalanche effect.

The avalanche effect is measured by the variation of output data, when the input data is altered. Hence, an encryption algorithm must show greater avalanche effect score, such that any alteration made can show significant changes. Suppose when a single bit is modified, considerable changes are observed. The avalanche score (AVS) ranges in the scale of 0 to 1 and is computed by the following formula.

$$AVS = C_{mb}/T_b \tag{3.3}$$

C_{mb} is the total count of modified bits in the cipher text and T_b is the total number of bits in the cipher text. When the AVS is greater, the encryption algorithm is proven to be stronger and efficient. The following Table 3.1 presents the AVS of different encryption algorithms. The cipher text is altered by a single bit, to carry out this analysis in different places (initial, middle, last).

Table 3.1: AVS of Different Encryption Algorithms

Place of Bit modification	Blowfish	DES	AES
Initial	0.008	0.43	**0.996**
Middle	0.007	0.46	**0.994**
Last	0.007	0.44	**0.996**

This analysis is carried out by varying the size of the data and the average AVS is presented in the above table. On observing the AVS, the AES algorithm is found to be a better performer than other algorithms and so AES is chosen for TSS. Additionally, the time consumption of AES is analysed and the results are as follows. This analysis is performed by varying the input size.

Table 3.2: Time Consumption Analysis of Encryption Algorithms

Size of input (kb)	Blowfish (ms)	DES (ms)	AES (ms)
10	4.8	8.3	**11.9**
39	8.8	33.2	**20.9**
56	16.2	53.6	**22.4**
112	26.8	67.4	**32.8**
168	36.7	83.1	**43.1**

The time complexity is measured in terms of milliseconds. Blowfish algorithm consumes minimal period of time, but it fails in exhibiting better AVS. The main intention of this work is to ensure better security rather than time conservation. Yet, the time consumption is tolerable. The time consumption of DES is greater, as the size of data increases.

AES shows stable performance. For this reason, AES is utilised in combination with Keccak. The AVS of the combination of Keccak with other algorithms are presented in the following table.

Table 3.3: AVS of the Combination of Encryption Algorithms

Place of Bit modification	Blowfish + Keccak	DES + Keccak	AES + Keccak
Initial	0.032	0.48	**0.998**
Middle	0.029	0.49	**0.997**
Last	0.03	0.47	**0.998**

Hence, the choice of AES among the encryption algorithms and the combination of Keccak with AES is justified. The proposed approach is evaluated by varying the encryption algorithms, SHA3 member algorithms. Finally, the Keccak is combined with different encryption algorithms and the results are discussed. From the experimental results, it is proven that the performance of the combination of AES with Keccak is proven with better results. The following section concludes the chapter.

3.4 CHAPTER SUMMARY

This chapter presents a security model for the cloud storage to preserve the data against numerous security threats. A twin layered security mechanism is used to preserve the integrity of the cloud storage. This is carried out with the Keccak's hash function and AES. The security of the system is tightened as the data integrity is preserved.

The security model is divided into two important phases namely hashing phase and encryption phase. In the hashing phase the data integrity is verified with the hash function which enhances the security of data. A tree is constructed with the hash codes and the computed hash codes are subjected to encryption with AES algorithm. As both hashing and encryption are present in the model, the security is guaranteed at the expense of computation. To check the integrity of data, the AES algorithm decrypts the hash codes which are verified with the original ones.

A sample work of TSS is discussed. Finally the performance analysis of the system is done in terms of memory consumption, execution speed and execution time. The system is compared with the SHA3 member algorithms and Merkle hash tree. The choice of AES is justified in terms of AVS when compared to the performance of Blowfish and DES. The system attains greater throughput and thus the execution time is minimized. Since blocks are used and hash codes are stored instead of the data, the memory consumption is much reduced. The time taken for execution is also lesser.

When Keccak and Merkle tree are compared, Keccak proves much lesser execution time as it is able to process multiple blocks of data at a time. Moreover, the storage overhead is minimal and the data transfer is proven to be better. The experimental results of the performance analysis of the twin layered security scheme show the efficacy of the work.

CHAPTER 4

SECURE STORAGE MODEL FOR CLOUD COMPUTING ENVIRONMENT BY EMPLOYING DISTRIBUTION AGENT

The introductory part of the chapter gives an overview and the motivation of the proposed Secure Storage Model in section 4.1. Section 4.2 presents the proposed SSM in which the assumptions, general flow, data block formation and distribution parts are discussed. The performance of the work is analysed in section 4.3 and section 4.4 concludes the chapter.

4.1 INTRODUCTION

Data is the prerequisite for any kind of task and thus, it is necessary to treasure the data. In today's world, all the activities generate useful data for future reference. However, it is a burden for the small and mid scale industries to store and manage the data. Cloud computing environment provides a solution to this issue by providing SaaS. The SaaS relieves the industries from the hardship of data management and overthrows the need for extra investment on storage equipments. However, the crucial point of data outsourcing is all about data security and privacy.

The CSP is the most important entity, which allocates services to the cloud users based on the demand and the clouds users have to pay only for the services they are utilizing. The resources of the cloud can be accessed from anywhere at any time.Cloud computing offers many services to the end users. Of these services one of the popular and most widely used services is cloud data storage. With the cloud data storage, the subscribers can store their data in the servers of the CSP instead of storing them in their own systems. There are many resources in the cloud. When the subscribers of the cloud use these resources or other services provided, they have to pay for the utilized services. This paradigm of cloud computing provides flexibility and scalability in data storage. The user need not have any concern about the storage mechanisms and issues regarding maintenance of the data storage. The data stored in the servers of the CSP can be accessed from any geographical area when the users stay connected to the cloud as cited in Reddy and Reddy [2012]

As far as SaaS isconcerned, the storage space is allocated for the cloud users with respect to the demand and SLA is signed between the cloud user and CSP.Inspite of all the merits of

SaaS, the data owners are reluctant to outsource their data, owing to data security. The data owners feel that the CSP may sniff the data being stored and may use it for their purpose.

As far as SaaS isconcerned, the storage space is allocated for the cloud users with respect to the demand and SLA is signed between the cloud user and CSP. Inspite of all the merits of SaaS, the data owners are reluctant to outsource their data, owing to data security. The data owners feel that the CSP may sniff the data being stored and may use it for their purpose. Thus, the main matters of concern in SaaS are data security and privacy, as cited by BNA [2009], Khan *et al.,* [2014*a*], Wang *et al.,* [2013].

Data security is the global term and there are so many security factors associated with it. Some of the important facets of data security are data integrity, confidentiality, availability, privacy and so on. Data integrity must be preserved by maintaining the data without any unwanted alterations. However, the data integrity may get affected due to the misbehaviour of the CSP.

It is difficult for the CSP to ensure data confidentiality, as numerous cloud data owners share the cloud server being provided by the CSP. Data availability is the most important parameter, which makes the service available round the clock. Besides this, the CSP must ensure the users that the data shared is safeguarded. The outsourced data must live even in the occurrence of any unfavourable natural calamity. Thus, it is necessary for the CSP to guarantee the availability of the service.

Taking all the security issues into account, this chapter intends to present a secured storage model for cloud computing that is capable of meeting the privacy and security requirements. The basic idea conceived in this work is not to employ multiple CSP, but to have multiple cloud servers of the same CSP. The proposed storage model utilizes multiple cloud servers to achieve the data integrity and availability. The main reason for the employment of multiple cloud servers is to weed out the total dependence on a single cloud server. This work splits up the data into several blocks and distributes the data blocks to different clouds. This way of data distribution after data split up improves the data integrity and availability. To achieve the goal, this work is subdivided into data block separation, encryption, data distribution and data retrieval.

In continuation to the previous work in the preceding chapter, the data block separation is performed by Keccak's hash function and the encryption is carried out by AES algorithm. The encrypted data blocks are distributed by the agent to different cloud servers by taking the current load and traffic of the server into account.

The flow of the data is marked in a separate file called catalogue file, which is maintained by the agent. This file is shared with the data owner, followed by which the catalogue file is deleted, in order to conserve memory. This provides a strong security as the proposed work sets several checkmates for the intruders. The following section explains the proposed secured storage model in detail.

4.2 PROPOSED SECURED STORAGE MODEL (SSM)

This section describes all the components of the proposed secure storage model in detail. Additionally, the assumptions along with the overall process flow of the work are presented.

4.2.1 Assumptions

The proposed work exploits the term 'multiple cloud server', which means that the cloud servers are from a single CSP. This work does not utilize multiple clouds, where each and every cloud belongs to different CSP. The word 'data' being used in this paper denotes the textual data and does not support any other format of data. This work utilizes the cloud space of multiple cloud servers of the same CSP.

The cloud servers are distributed across several geographical locations. The catalogue file is accessible to the data owner alone and it contains the geographical location of the particular data block. The data owner can gain access to the catalogue file only when the identity is proved to the agent. The general flow of this work is presented below.

4.2.2 General Flow of the Proposed SSM

Now-a-days, data outsourcing is an inevitable concept for any environment. Data outsourcing is the process of handing over the lump-sum data to the CSP, which yields many profits to the data owner. The data owner pays the CSP for the service usage and is freed from the data management activity.

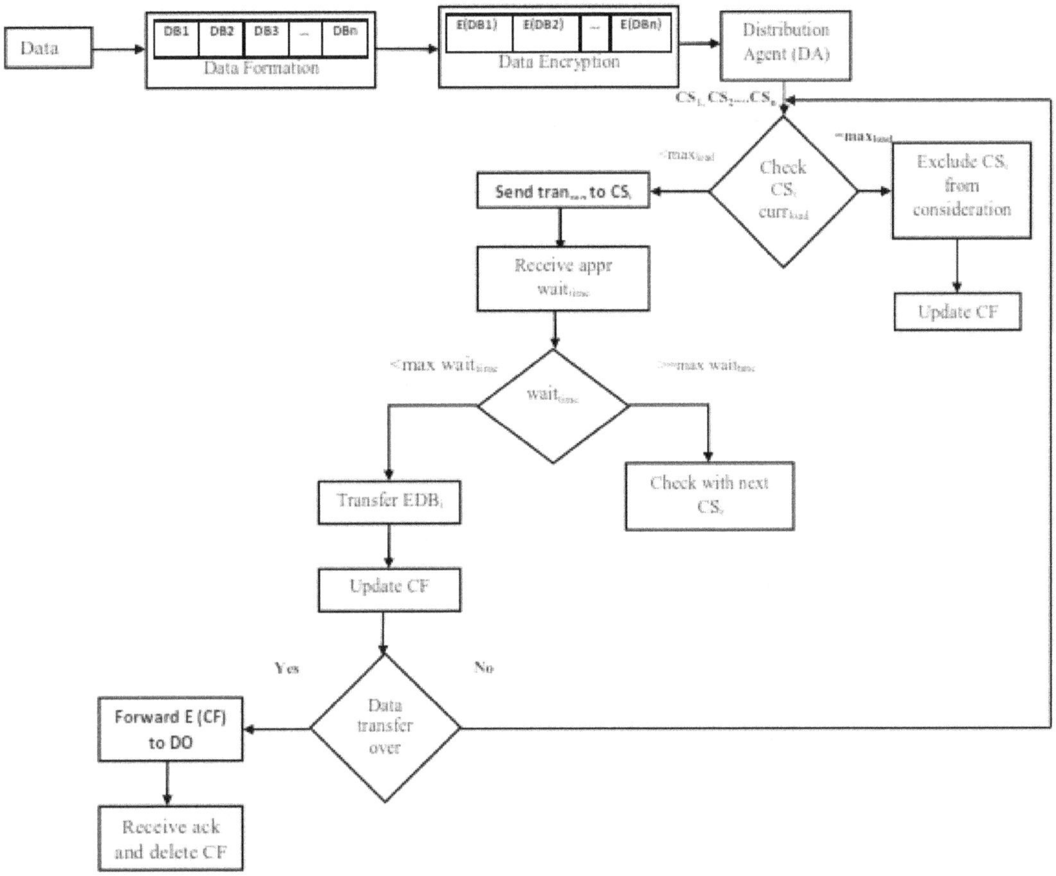

Figure 4.1: Flow of Encrypted Data Block Distribution By DA

However, the data owner is hesitant to hand over the confidential data to the external CSP with the concern that the CSP may misuse the data. Besides this, the data owner is also concerned about the integrity of the data, which means that the consistency of the data has to be maintained. In literature, so many works are proposed for preserving the data integrity.

This work is conceived on the idea that the integrity of the cloud data can be preserved, when the data is broken down into several data blocks and distributed to different cloud servers. Besides this, the hash codes of the data blocks are computed and encrypted as in the previous chapter. The encrypted data blocks are randomly distributed to multiple cloud servers, depending the data load and traffic.

This speeds up the process of distribution and improves the quality of service as well. The process of distribution is carried out by the distribution agent and it manages the catalogue file, which consists of the data blocks identifier along with the cloud server information. The overall flow of the proposed approach is depicted in Figure 4.1.

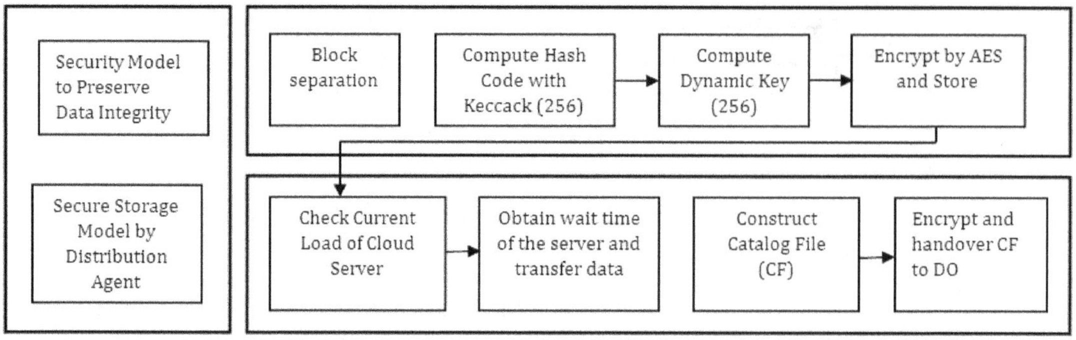

Figure 4.2: Layered Architecture of the Work

The catalogue file is shared only with the data owner with a simple security mechanism. The data owner is provided with the unique identifier and forwards the request for catalogue file to the distribution agent. The distribution agent verifies the identity and forwards the encrypted catalogue file. Symmetric key encryption is utilized to encrypt the catalogue file. The architecture is represented in layers in the following Figure 4.2.

Hence, this work is so conscious about the security of the cloud data. In fine, the cloud data is encrypted and distributed to several cloud servers, such that the intruders cannot guess the continuity of the data. The distribution agent maintains a catalogue file, which consists of the data block identifier along with the location.

4.2.3 Data Block Formation and Encryption

This work exploits TSS proposed in chapter 3. The TSS involves two phases, which are hashing and encryption phase. The hashing phase is achieved by means of keccak's hash function.

The algorithm for data block formation and encryption is presented below. The data to be outsourced is passed to the keccak's function, which decomposes the data into several data blocks. This is followed by the computation of the hash code by keccak's hash function. The algorithm for this phase is given below.

The encryption phase employs AES algorithm. At first, the dynamic key of size 256 bits is generated by CodeIgniter key generator. The hash code and the dynamic key are passed to the AES algorithm.

Algorithm 4 Data Block Formation and Encryption

Input: Data (D) to be outsourced, OL

Output: Encrypted data blocks (DB)

```
Begin
Break D into DBs w.r.t OL;
D = DB_1, DB_2, DB_3, ....DB_n
For DB_1 to DB_n
do
Compute HC;
Compute d;
Pass HC and d into AES;
Return cipher DB (EDB);
End;
End;
```

AES algorithm involves 256 bit key and 14 rounds for providing the cipher text. In the above given algorithm HC is the hash code and d is the dynamic key. The following section presents the scheme of data block distribution to multiple cloud servers

4.2.4 Data Block Distribution to Multiple Cloud Servers

As soon as the encryption process is over, the encrypted data blocks are distributed to different cloud servers by the distribution agent. The underlying ideas of distributing the encrypted data blocks are twofold. Initially, the intruders have to break the encryption algorithm to obtain the plain text.

Secondly, the intruders cannot follow the complete document, as the data is decomposed into several blocks and stored in several CSs. The existence of a specific data block cannot be guessed by the intruder. The main responsibility of the distribution agent is to allocate space for the encrypted data block.

The distribution agent maintains a table named cs_{detail} that includes the available cloud servers, geographical location along with the current load. The proposed algorithm for data block distribution is presented below.

Each and every cloud server is identified by a unique identifier and a CSP can have any number of cloud servers. This can be denoted as CSP=($cs_1, cs_2, cs_3, ...cs_n$). The cloud servers must be easily tracked by the distribution agent, so that the location of the cloud server is maintained.

Algorithm 5 Data Block Distribution by DA

Input: Encrypted data blocks;

Output: Distributed data blocks;

Begin
Instantiate DA to manage EDB and CS;
For EDB=1 to n
Do
Check cur_{load} of CS_i;
If $cur_{load} < max_{load}$
Do
Send $tran_{req}$ to CS_i;
Receive $tran_{res}$ from CS_i;
If ($wait_{time} > Max_{waittime}$)
Check with the next CS;
Else
Transfer EDB to CS_i;
End;
End;
For every successful EDB transfer
Update CF;
End;
End;
End;

Finally, the current load of the cloud server is retained, such that the cloud server will not be overcrowded. The distribution agent decides the location on which the data block has to be kept, by taking the load of the CS into account.

When the load of the CS is found to be minimal, then the distribution agent sends a transfer request $tran_{req}$ to the CS. The CS responds the distribution agent with $tran_{res}$ with the request queue number, if applicable. In case of the happening of multiple file transfer, then the distribution agent does not choose that particular agent.

Likewise, all the encrypted data blocks are transferred to the CS at different geographical locations. In this scenario, a new issue with respect to the traceability of the data block arises. For this purpose, a special file named 'catalogue' is maintained by the distribution agent. The importance of catalogue file is explained in the following section.

4.2.5 Importance of Catalogue File (CF) and its Security

The catalogue file stores the data block identifier along with the identifier of CS, in which it is stored. The catalogue file is maintained by the distribution agent till the last data block is placed in the CS. On the successful completion of data transfer process, the catalogue file is handed over to the data owner and the catalogue file is deleted from the local memory. This is to conserve the memory and also to ensure data privacy.

Besides this, it is absolutely necessary to provide security to the catalogue file, as it possesses the complete map of the data block existence. In order to ensure security, the catalogue file is encrypted by the symmetric key, which is generated by the distribution agent.

As soon as the data transfer process is over, the data owner is notified with the message that all the data blocks are successfully uploaded. On receiving this message, the data owner replies the distribution agent with the acc_{req} along with the do_{id} and timestamp ts. The distribution agent checks for the identity of the data owner and the encrypted catalogue file is forwarded to the data owner. The algorithm for this phase is given below.

Algorithm 6 CF Management by DA

Input: CF;

Output: Encrypted CF;

Begin

If data transfer completed

Do

Send message Upload success with ts to DO;

If acc_{req} message received

Encrypt CF with sym_{key} to obtain ECF;

Forward ECF to DO;

End;

If pos_{ack} received

Delete the ECF from local memory

If neg_{ack} received

Resend the ECF;

End;

End;

End;

The catalogue file (CF) is encrypted with the symmetric key. The communication between the distribution agent (DA) and the data owner (DO) is presented as follows.

$$DA \rightarrow DO : Upload\ success|ts \tag{4.1}$$

$$DO \rightarrow DA : acc_{req}|do_{id}|ts \tag{4.2}$$

$$DA \rightarrow DO : CFts_{(sym_{key})} \tag{4.3}$$

$$sym_{key} = \beta.(rand_{key}(DO_{id})) \tag{4.4}$$

$$DO \rightarrow DA : ackts \tag{4.5}$$

In the above equations, *DA* is the distribution agent, *DO* is the data owner, ts indicates the timestamp, acc_{req} is the access request, *CF* is the catalogue file, sym_{key} is the symmetric key, β is the pseudorandom prime number generator. $rand_{key}$ is generated by the distributed agent, which is of 128 bits.

Based on the *ack*, the catalogue file from the distribution agent is deleted. For instance, when the *ack* is negative, then the catalogue file is reforwarded to the data owner. On the contrary, when the *ack* is positive, the catalogue file is deleted immediately. The performance of the proposed work is evaluated in the following section.

4.3 RESULTS AND DISCUSSION

This work is simulated on Java platform in a machine with Intel i7 processor, 1.80 GHz and 16 GB RAM. The proposed work is subdivided into three different modules: (1) data block formation and encryption (2) data block distribution by DA and (3) Catalogue file management. As the performance of the first module is evaluated in our previous work, the remaining modules are tested for their efficiency.

The simulation is carried out ten different cloud servers and the data are distributed to them. The total data being processed by this work is 500 MB. The performance of this work is evaluated in terms of throughput, memory consumption, storage utilization, time consumption for data storage and data retrieval.

Throughput is the most important performance metric of any algorithm, as it decides the speed of the algorithm. The throughput is measured in Mbps (Mega bits per second). The memory consumption is another important requirement of any algorithm. The memory consumption must be preferably minimal, such that the proposed algorithm can be claimed as effective.

Storage utilization comes into picture, as this work distributes the encrypted data blocks by taking the current load of the cloud server. Thus, the storage utilization with respect to the CS is measured. It is also necessary to measure the time it takes to store and retrieve the data back from the CS.

The time consumption for data storage varies with respect to the total count of EDBs. The EDBs are varied from 5 to 20 data blocks. The total 500 MB data is divided into 5Œ100 MB, 10Œ50 MB and 20Œ25 MB. In order to check the performance, this work employs these five attributes and the results are presented through the figures 4.3-4.6.

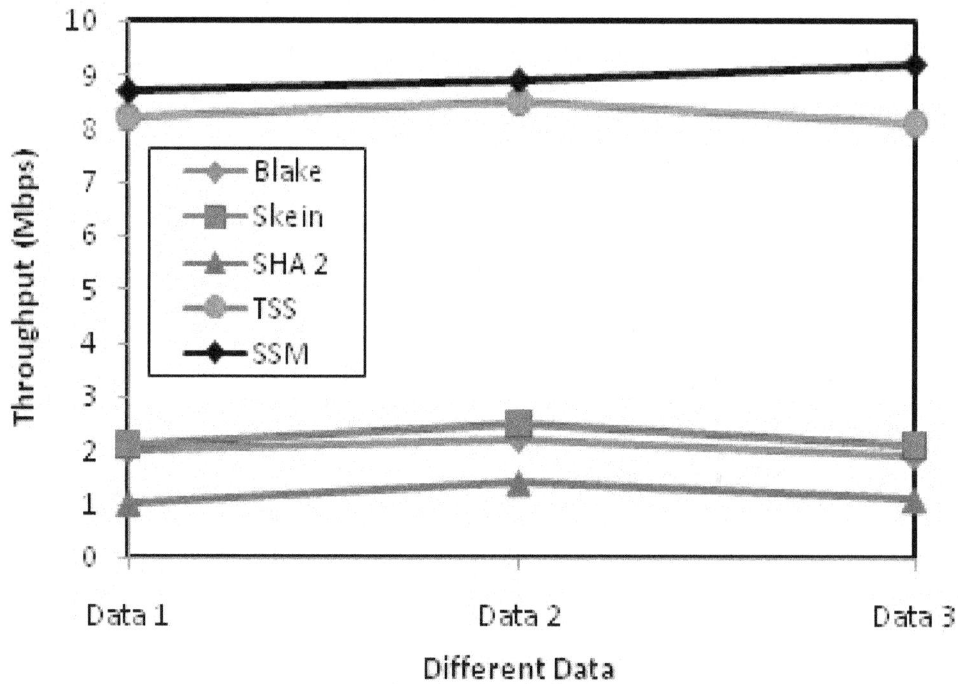

Figure 4.3: Throughput Analysis by Varying Data

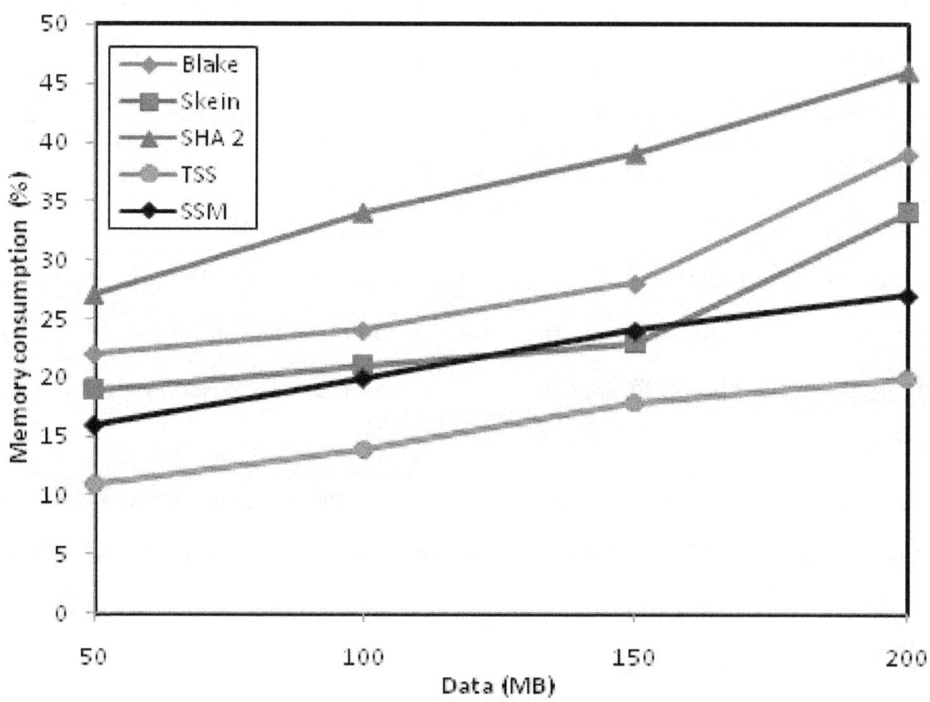

Figure 4.4: Memory Consumption Analysis

The throughput rate decides the speed of the algorithm and is measured in Mbps. This work varies the data to check the throughput and the performance is compared with the SHA

3 member algorithm discussed by Aumasson J.P. et al. (2010), Gauravaram P. et al. (2011), Guo X. et al. (2010). The throughput of SSM is slightly higher than the TSS, because of the impact of data distribution process.

The DA checks for the availability of the CS, before proceeding with the data transfer. The DA chooses the CS by taking the current load and traffic into account. In case, if the load or traffic is high for a particular CS, then the DA checks the next CS. This decision seriously impacts on the increase of throughput rate. Figure 4.4 shows the memory consumption of the proposed approach.

The memory consumption of the SSM is compared against SHA 3 member algorithms and our previous TSS. The experimental results show that the SSM consumes more memory than TSS. The reason for the increase in memory utilization is that the TSS does not store any metadata for future reference.

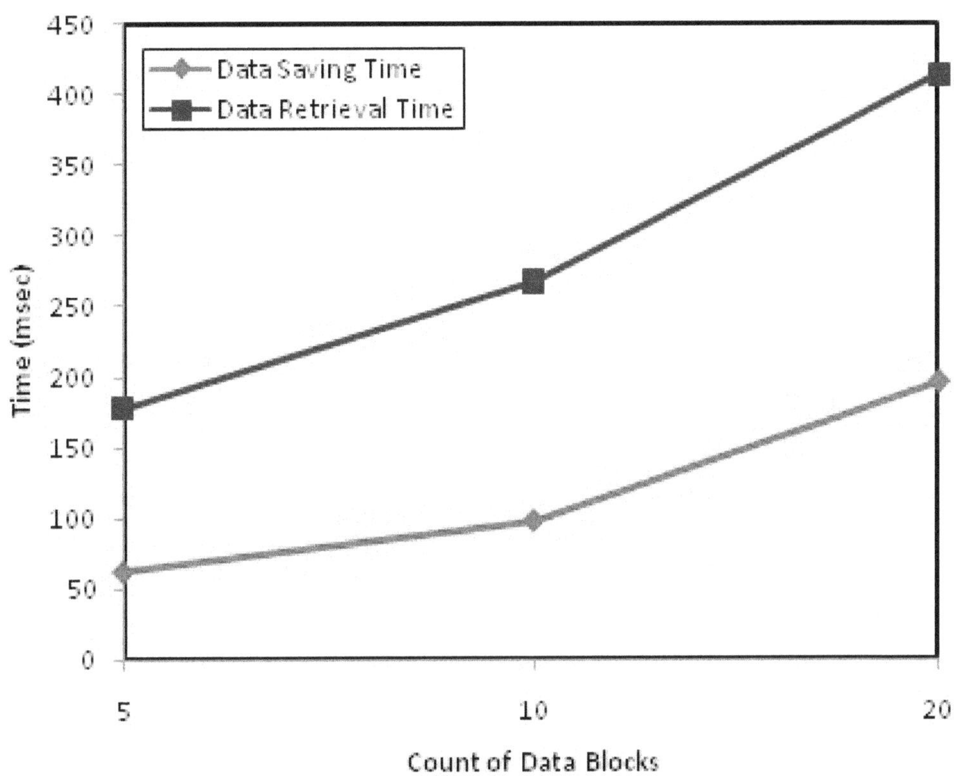

Figure 4.5: Analysis on Data Storage and Retrieval Time

Besides this, there is no involvement of any other party such as DA. On the other hand, SSM consumes more memory, owing to the employment of DA and the catalogue file. Though the catalogue file is deleted after its hand over to the data owner, the memory consumption improves. The data storage and retrieval time are measured and the results are presented in Figure 4.5.

The data saving and retrieval time for the complete 500 MB data are computed and the results are presented. Initially, the complete 500 MB is divided into 5Œ100 MB, 10Œ50 MB and 20Œ25 MB. The experimental results show that the data saving time is lesser, when compared to the data retrieval time. The reason for this is that the data are segregated and stored in different cloud servers.

The data retrieval time is measured by retrieving the complete data and thus, the data retrieval time is more. Besides this, while retrieving the data the sequence has to be maintained. On the other hand, when a particular data block alone is retrieved, the retrieval is done in a streak.

Both data storage and retrieval time increases with respect to the count of data blocks. As the count of data blocks increase, the work associated with it also increases. However, more data block ensures more security, even though the computational complexity is a bit high. Figure 4.6 depicts the storage exploitation of all the ten cloud servers.

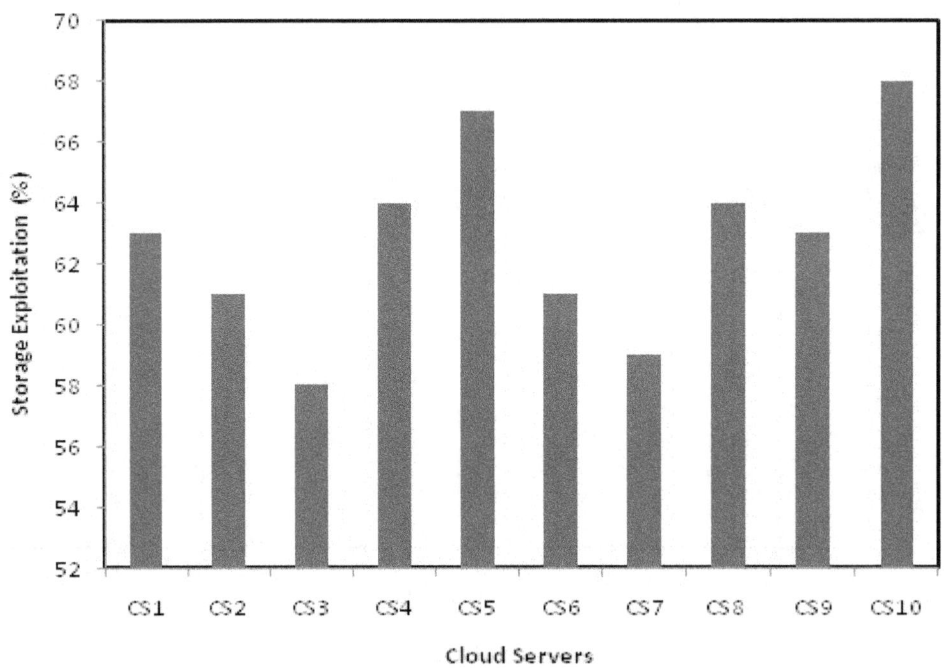

Figure 4.6: Storage Exploitation Analysis

The attained results are represented in tabular format, as in tables 4.1 to 4.4. The Table 4.1 shows the throughput analysis.

The results of storage exploitation analysis show the balanced distribution of data over

Table 4.1: Throughput Analysis

Data	Blake (Mbps)	Skein (Mbps)	SHA 2 (Mbps)	TSS (Mbps)	SSM (Mbps)
Data 1	2	2.1	1	8.2	8.7
Data 2	2.2	2.5	1.4	8.5	8.9
Data 3	1.9	2.1	1.1	8.1	9.2

Table 4.2: Memory Consumption Analysis (%)

Data / Techniques	**50**	**100**	**150**	**200**
Blake	22	24	28	39
Skein	19	21	23	34
SHA 2	27	34	39	46
TSS	11	14	18	20
SSM	16	20	24	27

Table 4.3: Data Storage and Retrieval Time Analysis

Data Blocks / Techniques	5	10	20
Data saving Time (msec)	62	98	196
Data retrieval Time (msec)	178	267	413

the clod servers. The entire data load is shared equally with all the cloud servers, based on the decision of the DA. The CS is chosen on the basis of the current load and traffic, which improves the quality of service and effective utilization of the storage space.

Table 4.4: Storage Exploitation Analysis

Cloud Server	Storage Exploitation (%)
CS1	63
CS2	61
CS3	58
CS4	64
CS5	67
CS6	61
CS7	59
CS8	64
CS9	63
CS10	68

From the experimental analysis, it is concluded that the proposed SSM ensures tightened security owing to the employment of data block separation and encryption, encrypted block distribution and the catalogue file management. The intruder has to break three levels of security to gain access to the original data, which is very tedious. The experimental results show satisfactory results and preserve the data integrity. The sample screens of this work are presented in the following figures.

The above presented Figure 4.7 shows the sample screen for data upload from the cloud data owner side. The data owner is prompted to upload any given file and the file is divided into data chunks. The server side is represented in the following figure.

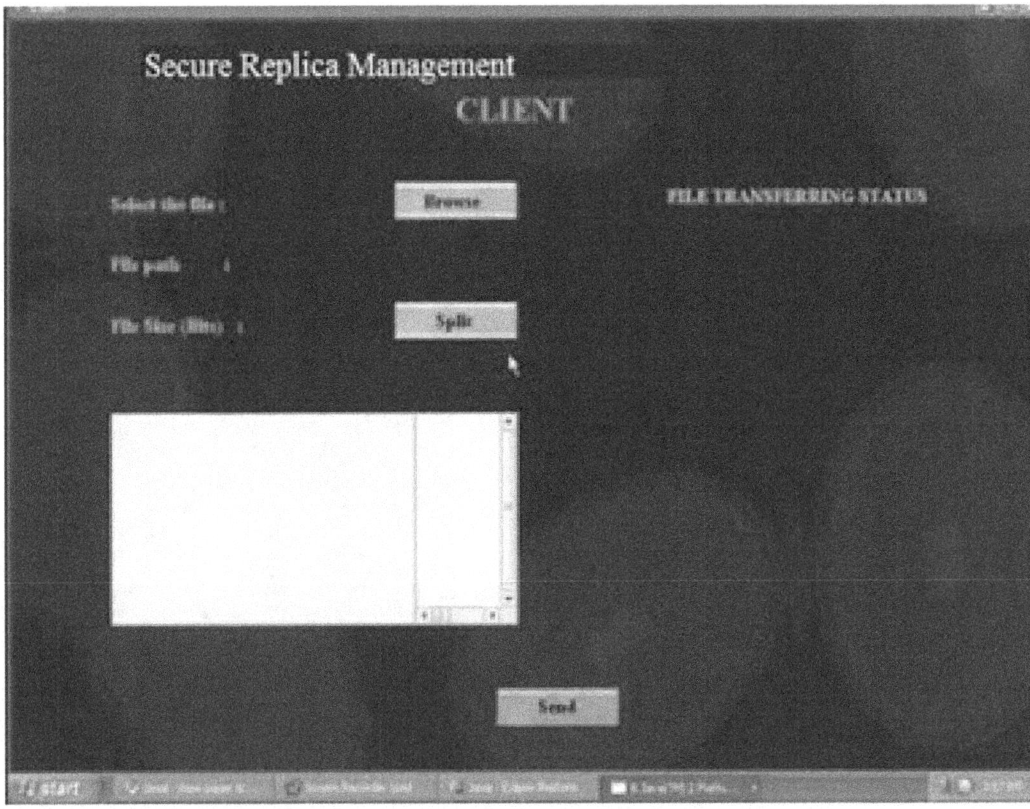

Figure 4.7: Sample Screen for Data Upload

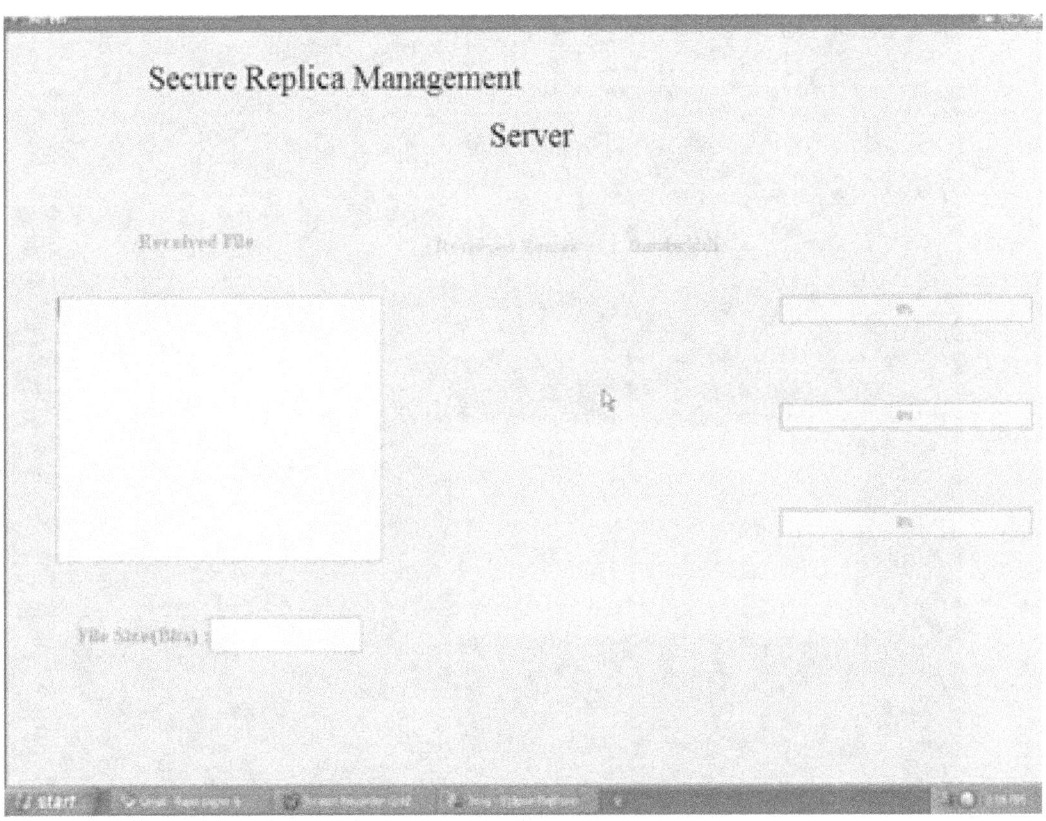

Figure 4.8: Server Side

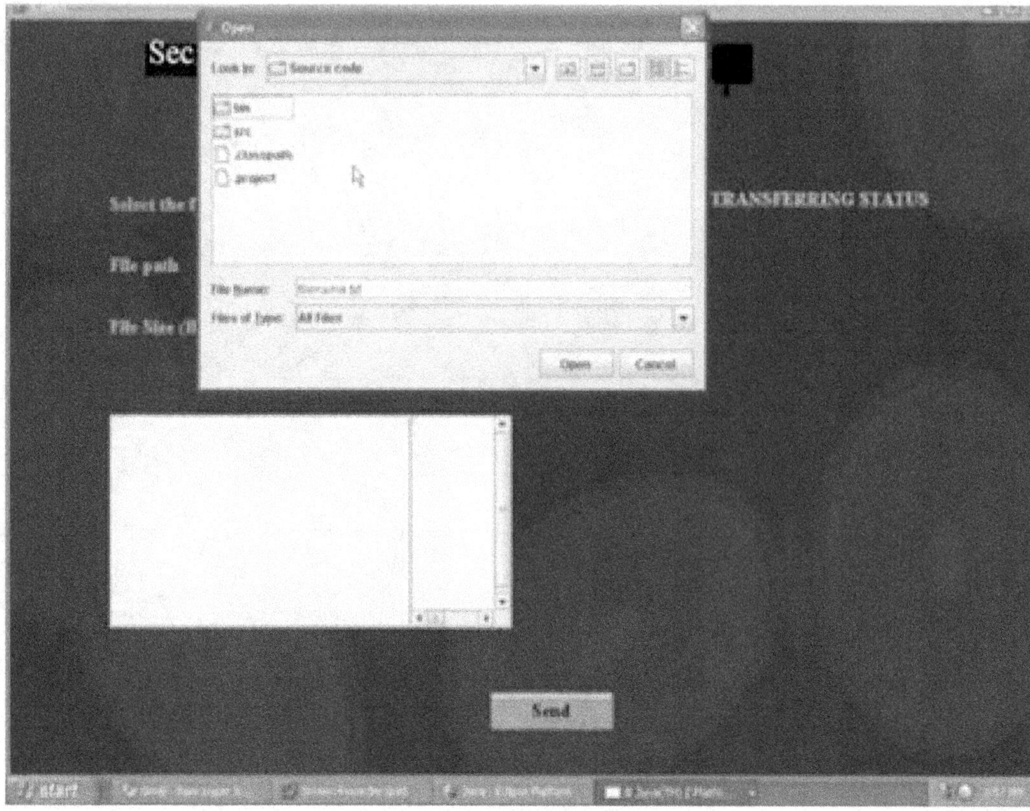

Figure 4.9: File Upload Browse

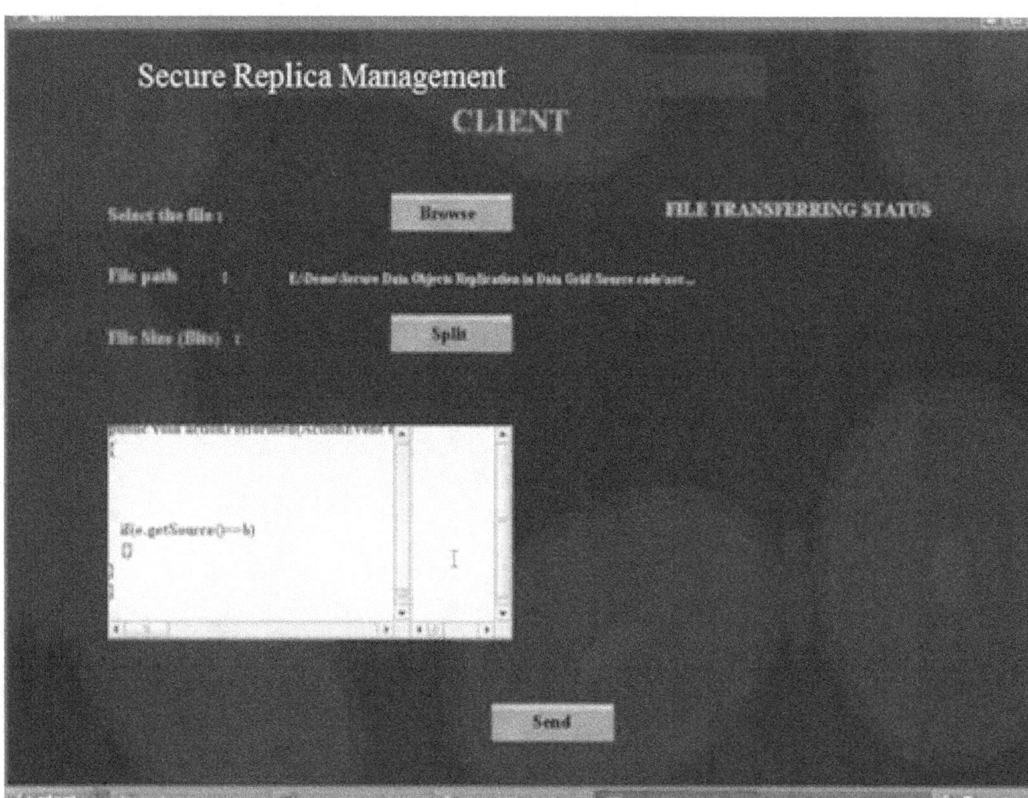

Figure 4.10: Uploaded File Screen

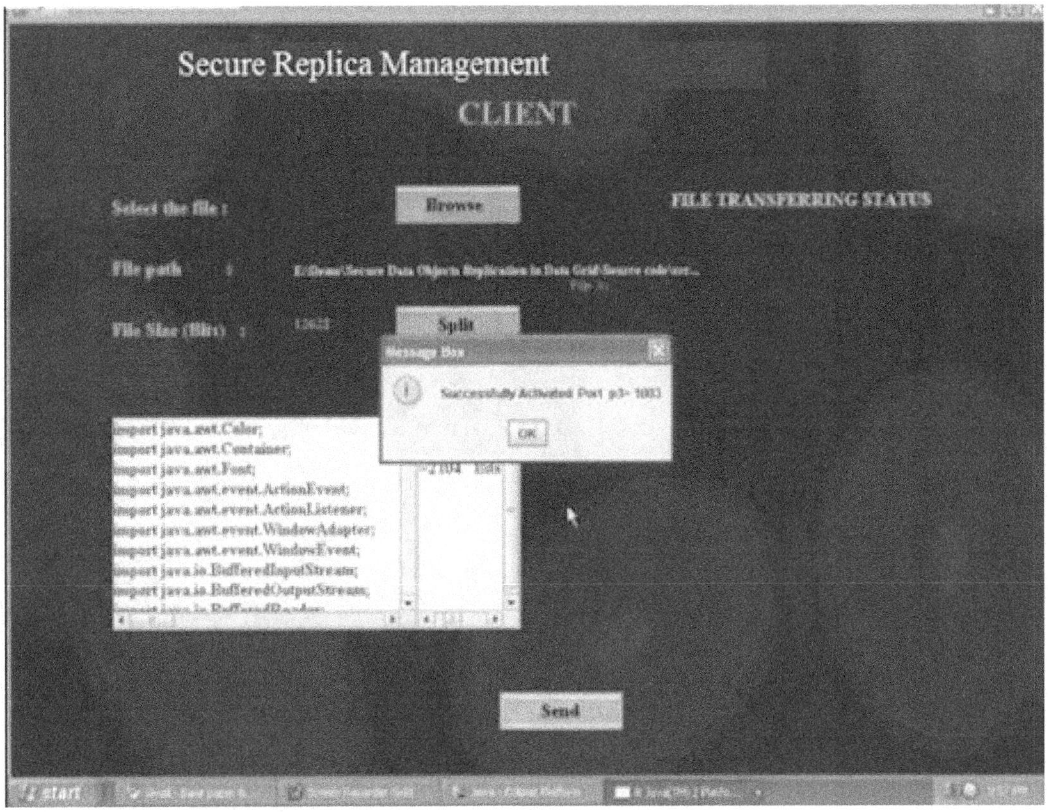

Figure 4.11: File Transfer Screen

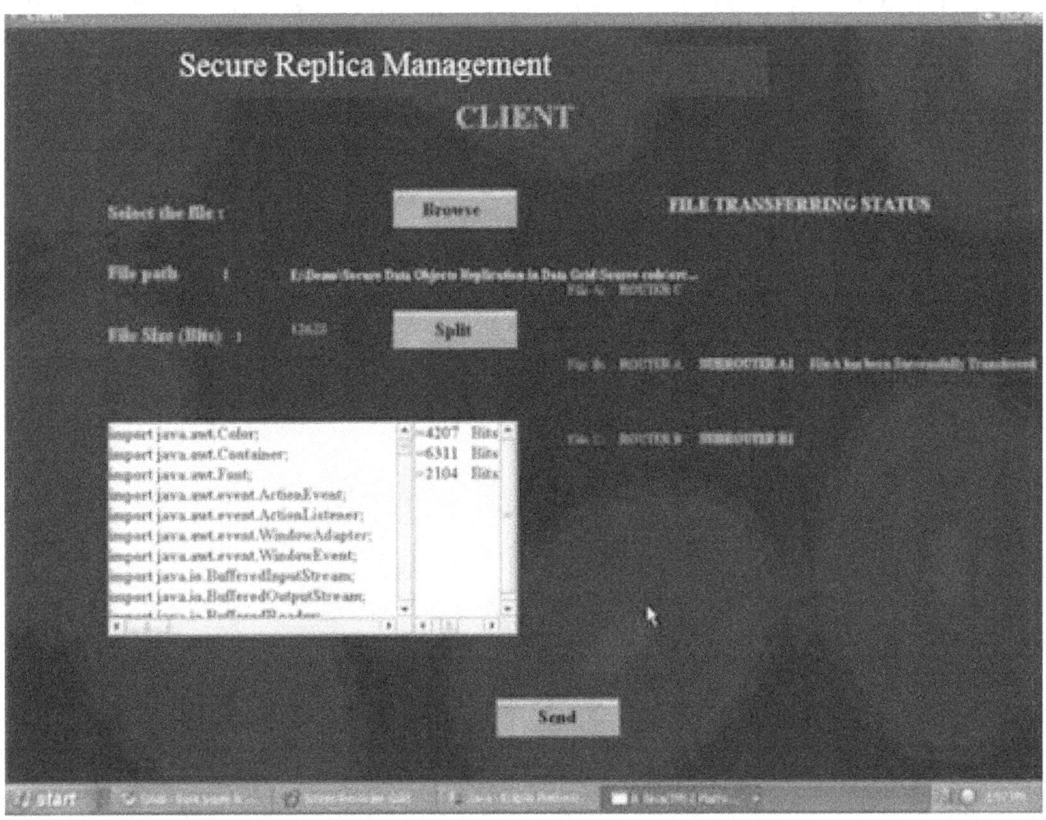

Figure 4.12: File Transfer Screen

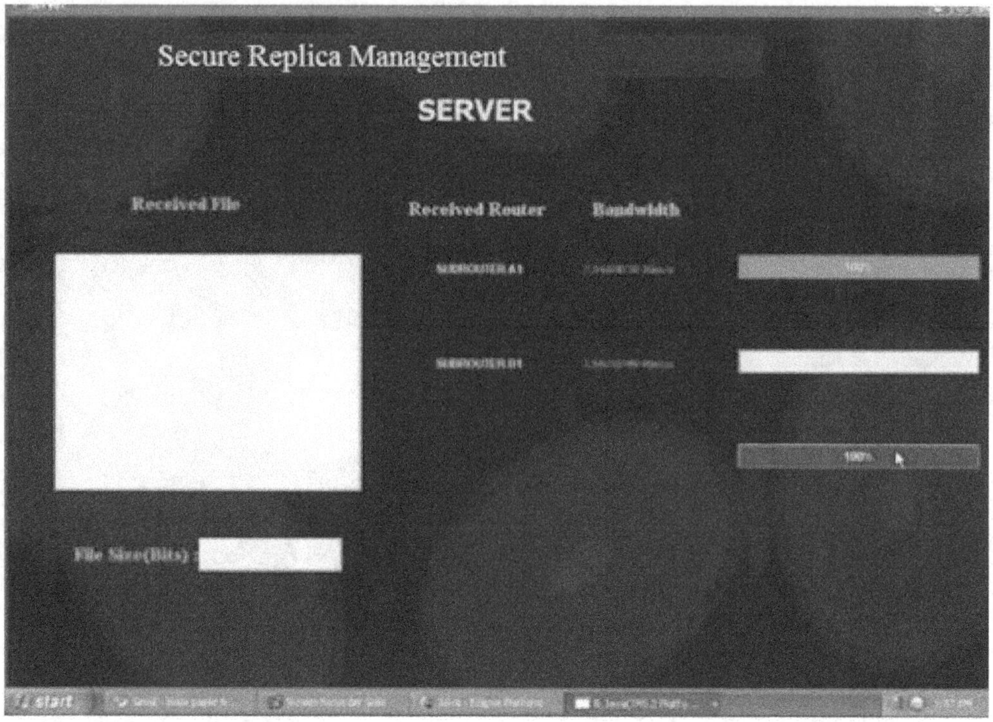

Figure 4.13: File Transfer Screen - Server Side

Hence, the sample screens of the work right from file upload is shared. The user can download the requested file at any instant of time and the sample screens for data download are shared.

Figure 4.14: File Download Screen

Hence, the proposed approach helps in storing the secured file, while making the data accessibility simpler. The summary of the chapter is presented as follows.

4.4 CHAPTER SUMMARY

This chapter presents a Secure Storage Model (SSM) for cloud computing, which intends to preserve the integrity of the cloud data. The SSM relies on three significant modules, which are data block separation and encryption; encrypted data block distribution by DA and catalogue file management.

This work introduces a distribution agent to take care of the distribution process of encrypted data blocks and the available cloud servers. The DA decides the cloud server on which the encrypted data block should reside. This decision is made by taking the current load and traffic of the cloud server into account.

The catalogue file is maintained by the DA till all the data blocks are successfully transmitted. The experimental results are satisfactory in terms of throughput, memory consumption and storage exploitation.

CHAPTER 5

SECURE DATA REPLICATION MANAGEMENT SCHEME WITH BETTER DATA ACCESSIBILITY

With the advancement in cloud computing, organizations are shifting the data from their storage to cloud to minimize cost and storage overhead, but face serious problems such as security and privacy. Security is associated with the factors such as confidentiality, integrity and availability. The TSS and SSM discussed in previous chapters solve confidentiality and integrity issues, but still data availability is a major concern. This chapter introduces the proposed Secure Data Replica Management Scheme (SDRMS) in section 5.1 along with its need. The proposed SDRMS is explained in section 5.2, which discusses the workflow, RMA's role, frequently accessed data detection and construction of tree-like structure and replica storage. The performance of the SDRMS is evaluated in section 5.3 and the chapter is concluded by section 5.4.

5.1 INTRODUCTION

Cloud computing is an inevitable technology of today's computing and storage environment, as it offers attractive service and deployment models. The Cloud Service Consumers (CSC) can choose any of the service and deployment models, based on the requirements. This flexibility results in the increased usage of cloud services.

The services being provided by the CSP are differentiated into platform, infrastructure, software and storage. The cloud service models encapsulate all the essential functionalities required by a business community. The business community can utilize all or any of the service models provided by the CSP, for which a reasonable payment has to be done. The deployment model is chosen based on the work environment of the CSC.

In spite of all the attractive service and deployment models, the cloud computing paradigm still suffers from certain shortcomings. The major concerns of cloud computing are security and privacy. As the private and sensitive data are shared with the CSP, it is necessary to ensure better security to the data. Privacy can be maintained while ensuring security. Though the CSP provides SLA to the CSC, the security breaches are still observed. Hence, cloud data security is the active research area and several algorithms are being proposed in the literature.

Data security involves securing data at storage and during transmission. The proposed work concentrates on data storage security.The parameters associated with security are data confidentiality, data integrity and data availability and are depicted in Figure 5.1

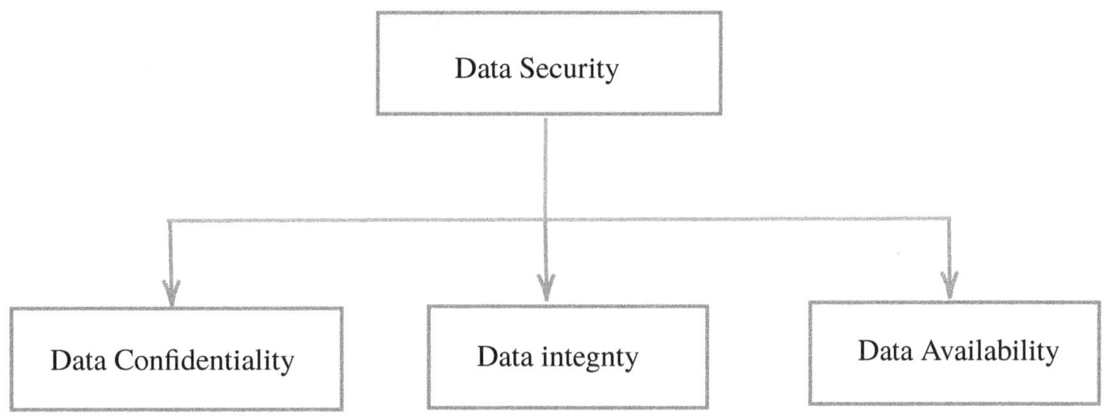

Figure 5.1: Organization of Data Security in Cloud Computing

Data confidentiality and data integrity issues are overcome by TSS and SSM and are discussed in last two chapters. Data availability is another challenge confronted by cloud computing technology. This term makes sense that the cloud data must be available to all the CSC, irrespective of the count of service requests. The Computer Security Alliance Group has listed data loss and leakage as one among the major threats in cloud computing. To overcome data loss, data should be made available at all times whenever needed.

In order to achieve data availability, the cloud data is replicated and stored. Replication is the method of sharing data to ensure data consistency. Creating one or more replicas provide business continuity. Replication copies the data to multiple places in order to reduce access time and bandwidth consumption. The replicas increase the data availability, parallelism and reduce network traffic. Keeping replicas in multiple places improves data availability, but increases cost, storage overhead and operational overhead . The major concern to be considered in keeping replicas is how to place these replicas effectively to minimize the above problems. As in Tabet *et al.,* [2017] many replication strategies aimed to answer the following questions.

- What data should be replicated?
- When should the data be replicated?
- Where the new replicas should be placed?

Gopinath and Sherly [2018] mentioned that many data replication techniques are available and can be classified in to two categories: static data replication and dynamic data replication.

In static data replication the number of data replicas to be created and the place to keep the replica is stated initially during cloud system setup. Dynamic data replication can adapt changes and automatically create and delete replicas according to changing environment.

Wei *et al.,* [2010*b*] proposed CDRM, a Cost-effective Dynamic Replication strategy which calculate the popularity of a data file to create replica for the data file. CDRM captures the relationship between data availability and replica number. The replica is placed in suitable storage by considering the capacity. CDRM provides cost effective data availability and improves performance by balancing load by maintaining a minimal number of replica.

Data replication is a necessary tool for managing the data effectively, however it introduces several challenges such as data modification and deletion, as cited by Roman *et al.,* [2013],JoSEP *et al.,* [2010],Al-Ayyoub *et al.,* [2015]. When the original cloud data is undergone some modifications, the same modifications have to be reflected on the replicas also, which is highly challenging. The modifications must reflect the replicas immediately, so as to escape from data inconsistency. These prompt modifications are possible only when the replicas are organised by some means. Considering these facts, this work attempts to organize the replicas in tree format by taking the load of the CS into account.

This work intends to present a replica management scheme that organizes the replica through which the data based operations are performed effortlessly. To our knowledge, most of the existing works does not organize the replicas and the data based operations are not carried out.

This work focuses on time and performance efficiency by creating replicas for the data with more demand. This idea saves space and time. In addition to this, the so created replicas are organised in a tree like format, which results in easy management of replicas while performing data based operations. The following section elaborates the proposed approach.

5.2 PROPOSED SECURE DATA REPLICATION MANAGEMENT SCHEME (SDRMS)

This section presents the proposed secure data replication management scheme in addition to the general flow of the work. The architecture is represented in layers in the following Figure 5.2.

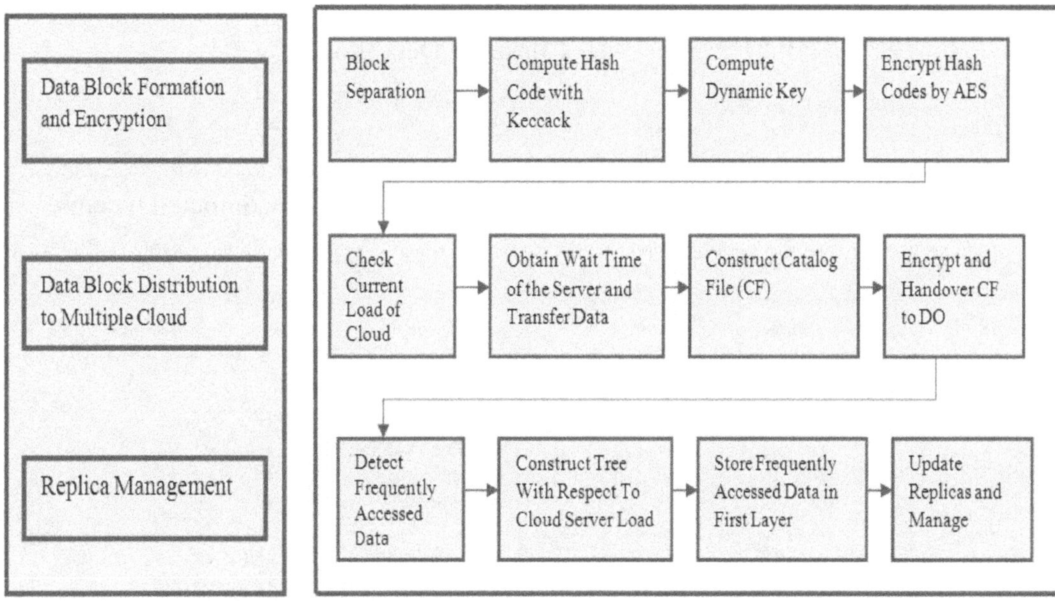

Figure 5.2: Layered Architecture of the Overall Proposed Work

As mentioned in previous chapters, data is divided into blocks of fixed size, hash codes for data blocks are computed and encrypted using AES algorithm. The encrypted data blocks are distributed to multiple cloud servers having minimal load and a catalog file is created for keeping track of data blocks placed in different servers. Inorder to maintain data availability, replicas are created for frequently accessed data blocks and are organized in a tree structure such that the frequently accessed data blocks get stored in the first layer and the updations are reflected immediately in the concerned data block.

5.2.1 Work flow of the Proposed Approach

Data is the pre-requisite of any task and hence, it has to be managed in a secure fashion. As all the simple process yields some beneficial data, it is difficult for the data owners to store and manage the available data. Additionally, the data is growing exponentially, which makes it tough for the data owners to provide room for it.

Cloud computing is a boon for the mid-scale industries, as it provides storage space for nominal charge. Though the cloud offers storage space to the data owners, the data owners hang-back to share the treasured data. Security is one of the major drawbacks of cloud, which makes the data owners to think thrice before outsourcing the data. However, the security aspect is presented in our previous works. The overall flow of this work is depicted in Figure 5.2.

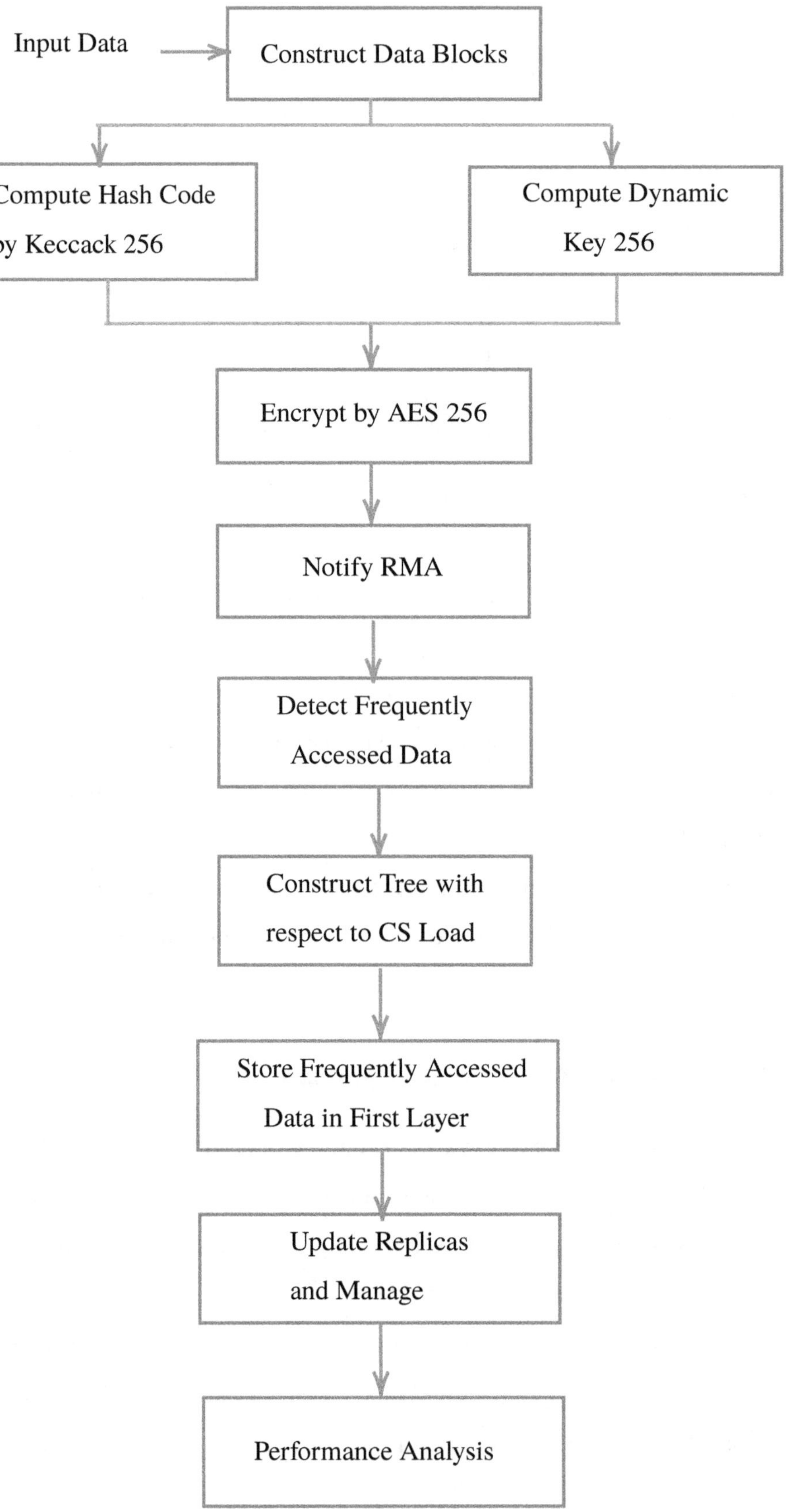

Figure 5.3: Work Flow of the Proposed SDRMS

Data availability is another crucial challenge of cloud computing, as it follows the distributed architecture. Hence in order to boost up the data availability, data replicas are created and managed. Though replicas are beneficial for increasing the data accessibility, it is expensive and is not easy to manage the replicas. The major difficulty in replica management is the replica upgradation. For instance, whenever the original data is modified or deleted, the data replica must reflect the changes made in the original data. Failing to do this, data inconsistency comes into picture.

On the contrary, it is not a good idea to create replicas for all the data unnecessarily. The reason is the unwanted memory allocation. Taking these issues into account, this work attempts to present SDRMS, which creates replicas in line with the access ratio of the data. Security is provided to the data replicas by means of TSS and the Replica Management Agent (RMA) takes care of the replicas.

The RMA requests for the current load of the servers and detects the data items with greater access rates. The RMA forms a tree like structure by taking the load into account. In each level of the tree, the cloud server is sorted in ascending order with respect to the load.

This makes sense that the first level of the tree contains cloud nodes with minimal load. Hence, all the replicas are loaded to the first level of nodes. Additionally, as the cloud servers are organized by some means, it is easy to update the replica as well. The overall work is shown in Figure 5.3. The locations of all the replicas and the cloud server are maintained by the RMA. The role of RMA is discussed in the following section.

5.2.2 Role of RMA

As the cloud follows distributed architecture, different RMAs are employed with respect to the geographical location. The role of RMA is shown in Figure 5.4.

Each RMA controls multiple CSs and the CS reports the RMA then and there. The RMA is responsible for creating replicas on the basis of the access ratio of the data.

In addition to this, the current load of the CS is checked periodically by the RMA. This periodical check-up is to avoid over-burdening the CS with more data. When the CS is overloaded with data, the performance of the CS goes down. Hence, the RMA considers the CS load for loading the replicas.

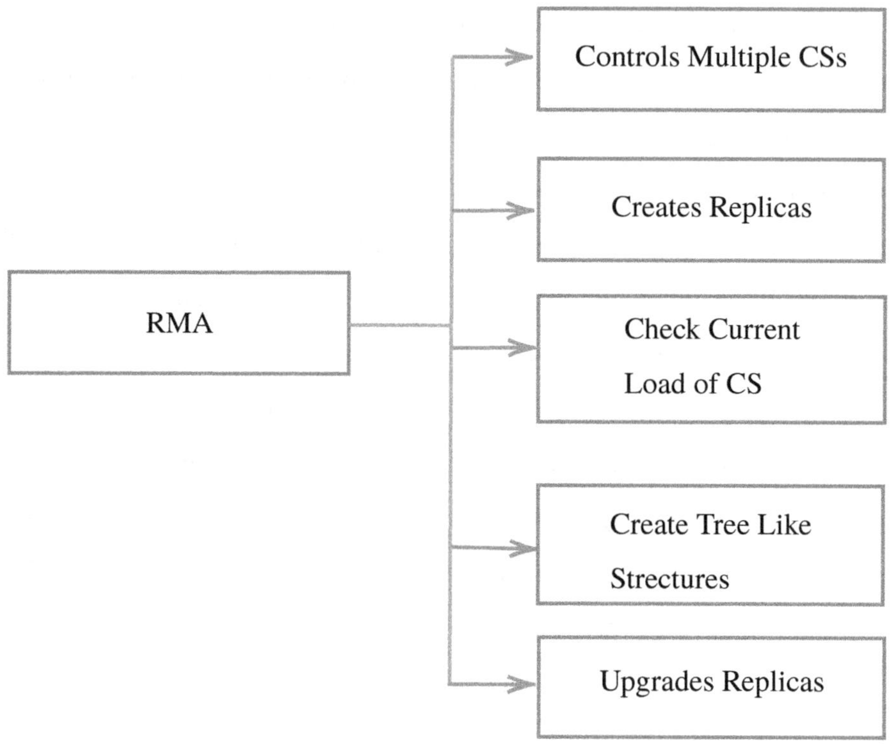

Figure 5.4: Role of RMA

On considering both the CS load and data access ratio, the replicas are created and loaded on to the CS. In addition to this, the RMA forms a tree-like structure, such that the first layer of the tree is loaded with the CS with minimal load.

Based on this principle, the replicas are preferably placed in the CS of the first layer. Hence, whenever the original data is modified, the RMA is reported. As the replicas are placed on the first level of the tree, it is easy for the RMA to upgrade the replicas. The following section presents the ways to detect frequently accessed data.

5.2.3 Detection of Frequently Accessed Data

The cloud contains voluminous data and it is not fair to create replicas for all the data, as all the data are not equally important. Replicas must be created for the important and frequently accessed data. This idea saves memory and increases the overall efficiency of the replica management system. The data accessing tendency of humans vary from time to time and hence, the RMA checks for the frequently accessed data on each CS for every period of time. In this case, the data access frequency checking interval for an RMA is set as 300 seconds. Here, there are two possible cases and they are change in access frequency and no-change in access frequency.

- Change in access frequency: In case, when some other data is accessed other than the previous frequently accessed data, the replicas are moved on to the second layer. Based on the access demand, the replicas are determined places.

- No-change in access frequency: In this case, the previous frequently accessed data maintains the same demand and it retains the same status with no change.

$$Freq_{adata} = \frac{(Request\ for\ data_i)}{(Total\ number\ of\ requests)} \tag{5.1}$$

When the $data_i$ is the most frequently accessed data out of all data requests, then obviously $data_i$ is the frequently accessed data. The value of $Freq_{adata}$ ranges from 0 to 1 and this Freq $_{adata}$ is arranged in descending order, such that the frequently accessed data comes first. The RMA requests for $Freq_{adata}$ and the current CS load for constructing the tree like structure and is explained in the following sub-section.

5.2.4 Construction of Tree-like Structure and Replica Storage

Based on the information that includes CS load and access frequency of the data, a tree-like structure is built by the RMA. This idea enables the process of loading the replicas easily and is makes the process of up-gradation easier too. The below presented Figure 5.2 presents the replica placement strategy. The overall algorithm of this work is as follows.

Algorithm 7 Proposed Algorithm for SDRMS

Input

1: Begin
2: Perform data encryption by TSS;
3: Notify RMA;
4: Detect frequently access data;
5: Construct tree like structure w.r.t load of CS;
6: Store frequently accessed data in the first level of tree;
7: Manage replicas;
8: **End**;

In Figure 5.4, N1 is the RMA. As presented in the figure, it is quite reasonable for any cloud server to be loaded with more data. On receiving current load from the CS, the RMA arranges the load in ascending order, such that the CS with minimal load comes first.

It is obvious that the CSs with minimal load are minimal for sure and these CS are given place in the first layer. Similarly based on the current load of the CS, all the CSs are given place in the tree-like structure. As observed in Figure 5.4, as the count of layers increase the CS with maximal load are present and is a normal case in cloud.

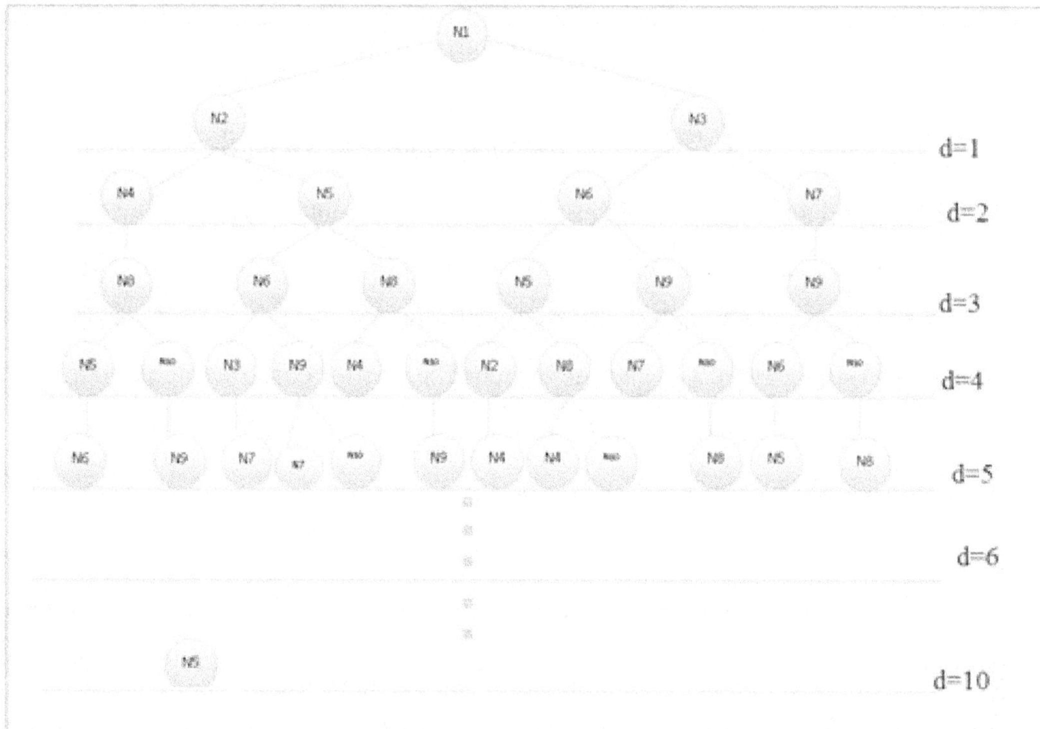

Figure 5.5: Sample Replica Placement Strategy

As soon as the tree is constructed, the replicas are loaded on to the CSs present in the first layer. Suppose after loading so many replicas, when the load of CS in the first layer becomes greater than the load of CS in the second layer, both the CSs are swapped.

By this way, the replicas are placed on the CS. Once the original data is upgraded, the replicas can easily be located and upgraded as well. The RMA stores the location of replica as well for easy up-gradation in a separate file, which is conceived on the basis of the catalogue file. The performance of the proposed work is evaluated in the following section.

5.3 RESULTS AND DISCUSSION

The performance of this work is tested and the results of this work are compared with the analogous techniques, discussed by Reddy *et al.,* [2017] and Liu *et al.,* [2014*a*]. The replica allocation system for cloud computing is scarce in the existing literature. This work provides both security and the replica management, which is an added advantage. The performance of this work is evaluated in terms of data accessibility, replica allocation time, replica retrieval time and storage exploitation analysis.

$$Data_{accessibility} = \frac{(Total\ number\ of\ service\ grants)}{(Total\ number\ of\ service\ requests)} \tag{5.2}$$

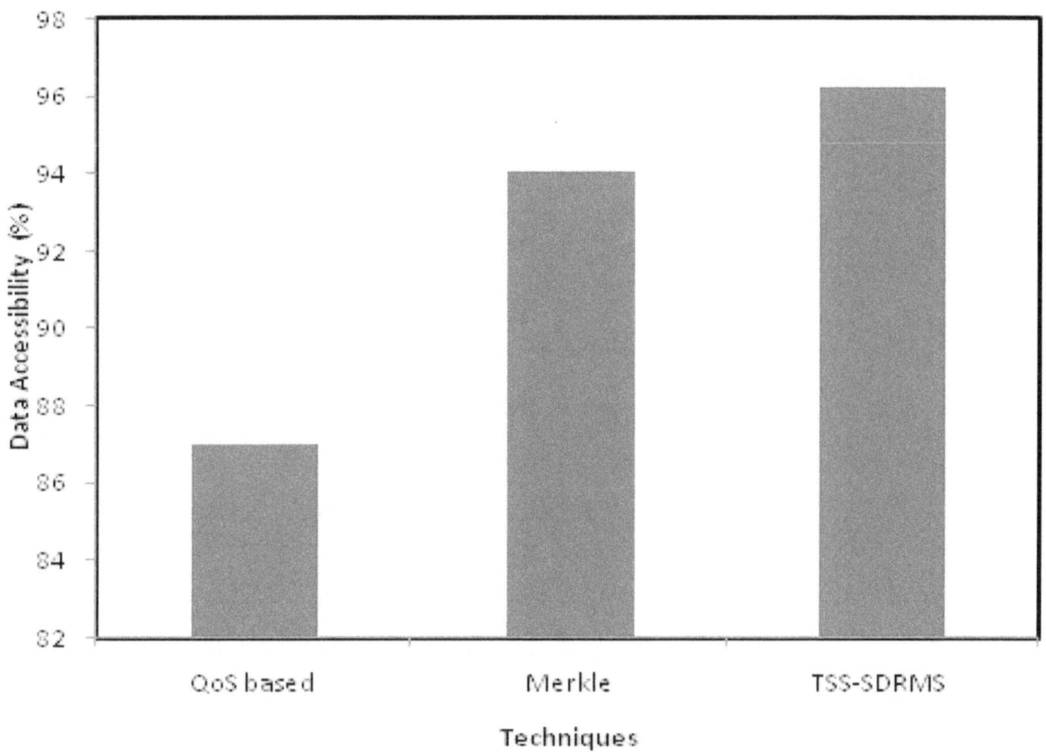

Figure 5.6: Data Accessibility Analysis

Better data accessibility is the most important objective of creating replicas. Creation of replicas helps in accessing the required data in a reasonable amount of time and this functionality has to be checked in the replica based schemes. Data accessibility is checked by considering the total count service grants to the total number of service requests and the performance of the proposed approach is analysed with the existing approaches.

On analysis, it is observed that the data accessibility of the proposed approach is better with greater data accessibility rate. The main reason for the improved data accessibility of the proposed work is the core idea of replica creation based on the demand of the data. In addition to this, the replicas are organised which helps in achieving easy access to the data.

On the other hand, QoS based approach employs MCMF pattern, which results in increased time complexity due to the poor data accessibility. The merkle based technique employs tree like format for each replica, which leads to computational complexity and reduces the data accessibility rate. The following graph presents the replica allocation time.

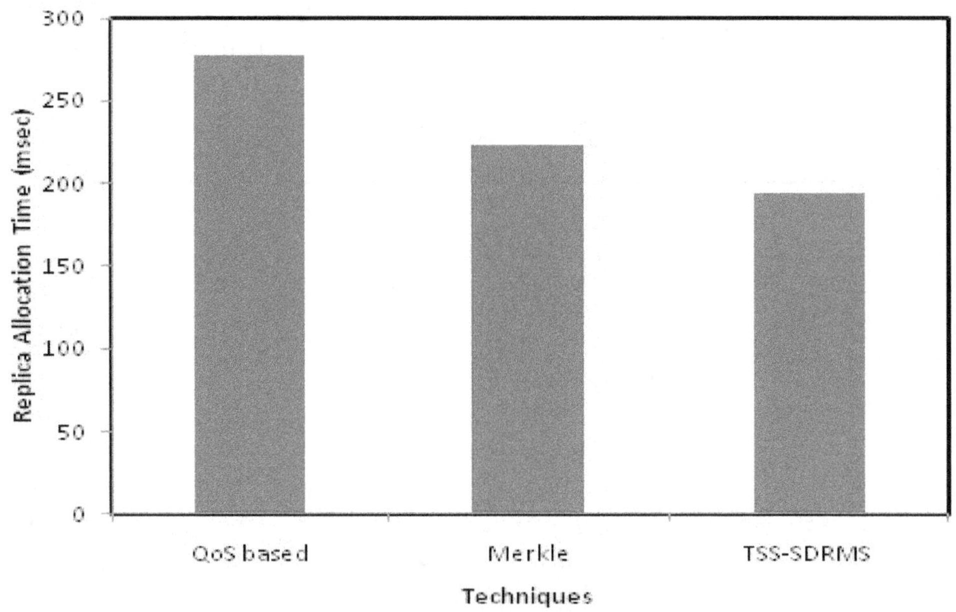

Figure 5.7: Replica Allocation Time Analysis

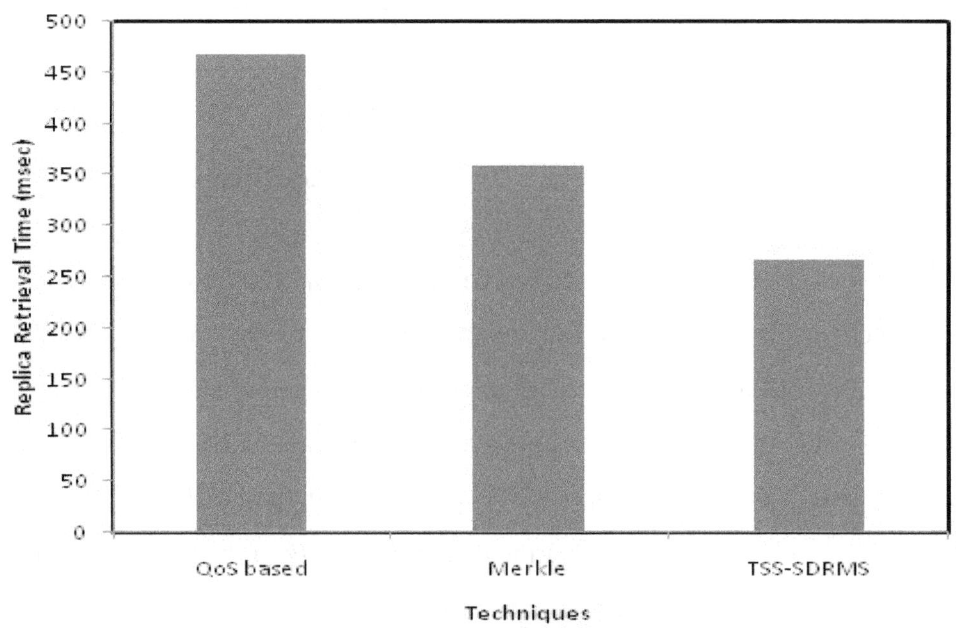

Figure 5.8: Replica Retrieval Time Analysis

The replica allocation time is computed by considering the creation of replicas and the allotment to the service requests is measured. On analysis, the proposed approach consumes minimal time for allotting the replicas. The main reason for the minimal time consumption is that the proposed approach does not create replicas for all the data, but the data with greater access frequency.

This idea supports in allotting the replica faster than the existing approaches. The QoS based approach does not include any replica organization technique and the merkle approach forms a tree for all the data, which leads to complexity in replica allocation. As the proposed approach employs RMA based on location, the tree can easily be accessed and allotted.

The replica retrieval time denotes the time it takes to retrieve the replica from the cloud server. Obviously, the QoS based approach consumes more time, as it does not follow any replica organization technique for storing replicas. On the other hand, merkle approach follows tree like structure and the sub-tree is meant for every single replica. This makes the replica retrieval complex and the proposed approach maintains the frequently accessed replicas on the first level of the tree. This idea helps in easy localization of the replicas and hence can be retrieved in no time. The following graph presents the storage exploitation analysis of the proposed approach.

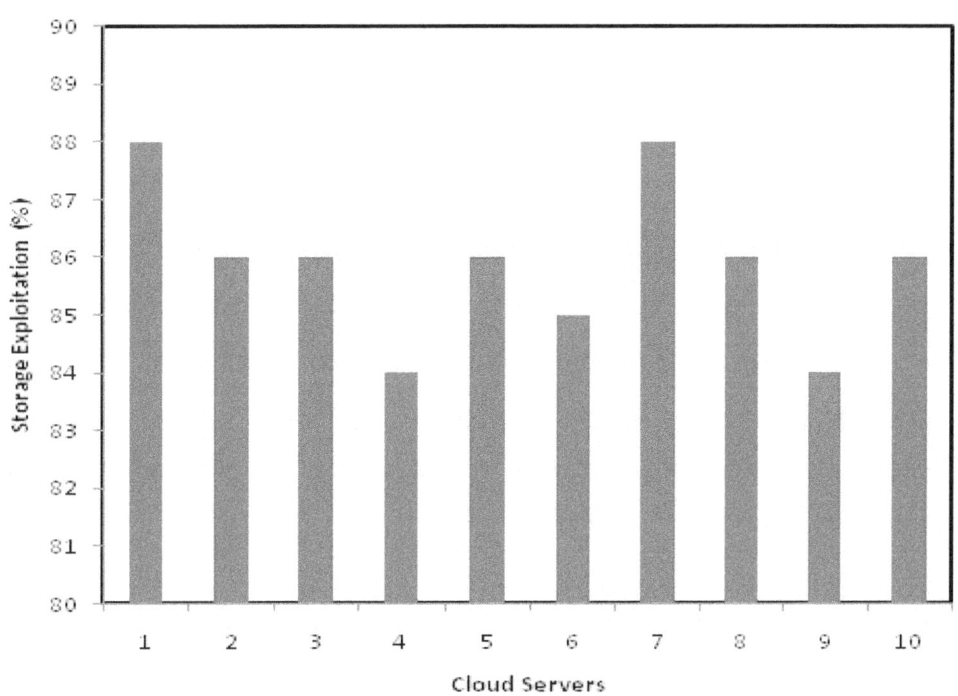

Figure 5.9: Storage Exploitation Analysis

Analysis on storage exploitation is one of the most important factors of any replica management system. The reason behind the fact is that no single cloud server should be overburdened by placing more number of replicas. The achieved experimental results are tabulated in tables 5.1 and 5.2 as follows.

Table 5.1: Experimental Results

Perf.metrics / Techniques	QoS based	Merkle	TSS - SDRMS
Data accessibility (%)	87	94	96.2
Replica allocation time (ms)	278	223	194
Replica retrieval time (ms)	468	359	267

Table 5.2: Storage Exploitation Analysis

Cloud Server	Storage Exploitation (%)
CS1	88
CS2	86
CS3	86
CS4	84
CS5	86
CS6	85
CS7	88
CS8	86
CS9	84
CS10	86

The proposed approach is much concerned about achieving better balance of load and hence, the cloud servers are sorted on the basis of load after which the replicas are placed. On analysis, it is noted that all the cloud servers are loaded in a balanced fashion and not overburdened. Hence, the proposed approach proves better results in terms of storage exploitation, data accessibility, replica retrieval and allocation time. The chapter is summarized in the following section.

5.4 CHAPTER SUMMARY

This chapter proposes a SDRMS for cloud data, which relies on TSS and RMA. The TSS is meant for encrypting the data and the replica management agent creates replicas based on the access frequency of the data.

The replicas are stored in the cloud server by considering the current load of the server. This idea conserves space, as the replicas are not created for all the data. In addition to this, the proposed approach achieves better load balance, such that all the cloud servers are utilized effectively.

The replicas are organised in a tree like structure and the frequently accessed replicas are placed on the first level of the tree, which increases the ease of data retrieval.

When compared with the existing approaches, it is concluded that the proposed method is efficient in terms of data accessibility, replica allocation time, replica retrieval time and storage exploitation analysis.

CHAPTER 6

CONCLUSIONS AND FUTURE ENHANCEMENTS

This chapter intends to share the conclusive ideas and the possible future enhancements of the research work are then presented. Additionally, the contributions of the research are also listed.

6.1 CONCLUSIONS

Cloud computing has gained substantial research interest, owing to its numerous advantages such as on-demand service, elasticity, pay for the exploited services and hassle-free work experience. Hence, many industries outsource their data to the cloud, in order to save memory and thereby cost. The CSP takes the responsibility of managing the data effectively. The main issue associated with the aforementioned point is the data security. The data owners are always concerned with the security of the data, that the CSP may misuse the stored data.

Data protection and security are the critical issues to be handled in cloud computing however, accomplishing the issues is not easy. As far as data security is concerned, data confidentiality, data integrity, authentication are its facets. When the confidential data are outsourced, data availability is another issue to be managed. The data availability can be ensured when the data stored in the cloud are accessible at all times.

Understanding these requirements, this research work attempts to focus on two major data security issues such as data integrity and availability. The central goal of the work is met by proposing three different security solutions for addressing the issues.

The initial research phase presents a data integrity preservation scheme for cloud data by presenting TSS. This work preserves the cloud data integrity in two steps and they are data hashing and encryption. Initially, the input data is decomposed into several data blocks and the hash code is computed for each block by employing keccak 256 algorithm.

A dynamic key of size 256 is then formed and the data encryption is carried out by AES algorithm. This idea of twin layered security ensures better security and the efficiency is proven with the help of avalanche effect and several other performance measures.

The second phase of research presents a secure storage model for cloud computing, which follows divide and conquer strategy in splitting the cloud data. On successful data split up, the data blocks are encrypted and stored randomly in various cloud servers, which exists in different geographical locations.

This way of dispersing the encrypted data blocks across multiple cloud servers tightens the security of the cloud data. Besides this, whenever the data is to be altered or deleted, the concerned data block alone can be extracted for manipulation. This conserves time and overthrows the computational complexity. Hence, this work is an improvisation of the previous work.

The final phase of this work proposes a replica management scheme that organizes the replica through which the data based operations are performed effortlessly. To our knowledge, most of the existing works does not organize the replicas and the data based operations are not carried out.

This work focuses on time and performance efficiency by creating replicas for the data with more demand. This idea saves space and time. In addition to this, the so created replicas are organised in a tree like format, which results in easy management of replicas while performing data based operations. This work ensures data availability with better security.

The performances of all the proposed approaches are tested in terms of standard performance measures such as memory consumption, execution time, throughput, data storage time, data retrieval time, storage exploitation analysis, data accessibility, replica allocation time, replica retrieval time and so on. The performances of the proposed approaches are compared with the existing approaches and the efficiency of the works are proven.

6.2 RESEARCH CONTRIBUTIONS

Some of the remarkable research contributions of this work are listed as follows.

- A data integrity preservation scheme called TSS is presented by combining the Keccak and AES algorithms for the cloud data.
- A secure storage model for cloud computing is presented for preserving the integrity of the cloud data by introducing a distribution agent in the system.
- A secure data replication management scheme is proposed for achieving both the data integrity and availability of the cloud data.
- The security of the cloud data is enhanced at the cost of time consumption.

- Data storage and retrieval time are considerably reduced.

Hence, this section presents the contributions of the research and the following section presents the possible future enhancements of the research work.

6.3 FUTURE ENHANCEMENTS

The possible future enhancements of this research work are listed as follows.

- The concept of multi-clouds with data replicas can be implemented, as this work considers multiple cloud servers but not multiple cloud providers.
- The data storage and retrieval time can be even minimized by effective data organization.
- This work can be extended to consider a real-time scenario for attaining better data availability and security as well.

REFERENCES

1. **Ahmed M.**, **Xiang Y.**, and **Ali S.**, Above the trust and security in cloud computing: a notion towards innovation, *In IEEE/IFIP International Conference on Embedded and Ubiquitous Computing*. IEEE, 2010, 723–730.

2. **Al-Ayyoub M.**, **Jararweh Y.**, **Daraghmeh M.**, and **Althebyan Q.**, (2015), Multi-agent based dynamic resource provisioning and monitoring for cloud computing systems infrastructure, *Cluster Computing*, **18**(2), 919–932.

3. **Ali M.**, **Khan S. U.**, and **Vasilakos A. V.**, (2015), Security in cloud computing: Opportunities and challenges, *Information sciences*, **305**, 357–383.

4. **AlZain M. A.**, **Pardede E.**, **Soh B.**, and **Thom J. A.**, Cloud computing security: from single to multi-clouds, *In 45th Hawaii International Conference on System Sciences*. IEEE, 2012, 5490–5499.

5. **AlZain M. A.**, **Soh B.**, and **Pardede E.**, Mcdb: using multi-clouds to ensure security in cloud computing, *In IEEE Ninth International Conference on Dependable, Autonomic and Secure Computing*. IEEE, 2011, 784–791.

6. **Amazon E.**, (2009), Amazon elastic compute cloud, *Retrieved Feb*, **10**.

7. **Aral A.** and **Ovatman T.**, (2018), A decentralized replica placement algorithm for edge computing, *IEEE Transactions on Network and Service Management*, **15**(2), 516–529.

8. **Armerding T.**, (2012), The 15 worst data security breaches of the 21st Century, *COS Security and Risk.*

9. **Attanasio C.**, Virtual machines and data security, *In Proceedings of the workshop on virtual computer systems*. ACM, 1973, 206–209.

10. **Aumasson J.-P.** and **Bernstein D. J.**, SipHash: a fast short-input PRF, *In International Conference on Cryptology in India*. Springer, 2012, 489–508.

11. **Ba-Hutair M. N.** and **Kamel I.**, A new scheme for protecting the privacy and integrity of spatial data on the cloud, *In IEEE Second International Conference on Multimedia Big Data (BigMM)*. IEEE, 2016, 394–397.

12. **Bernstein D.** and **Vij D.**, Intercloud security considerations, *In IEEE Second International Conference on Cloud Computing Technology and Science*. IEEE, 2010, 537–544.

13. **Bhadauria R.**, **Chaki R.**, **Chaki N.**, and **Sanyal S.**, (2011), A survey on security issues in cloud computing, *IEEE Communications Surveys and Tutorials*, 1–15.

14. **BNA P.**, (2009), Privacy & security law report, .

15. **Bohn R. B.**, **Messina J.**, **Liu F.**, **Tong J.**, and **Mao J.**, NIST cloud computing reference architecture, *In IEEE World Congress on Services*. IEEE, 2011, 594–596.

16. **Buyya R.**, **Yeo C. S.**, and **Venugopal S.**, Market-oriented cloud computing: Vision, hype, and reality for delivering it services as computing utilities, *In 10th IEEE International Conference on High Performance Computing and Communications*. Ieee, 2008, 5–13.

17. **Buyya R.**, **Yeo C. S.**, **Venugopal S.**, **Broberg J.**, and **Brandic I.**, (2009), Cloud computing and emerging IT platforms: Vision, hype, and reality for delivering computing as the 5th utility, *Future Generation computer systems*, **25**(6), 599–616.

18. **Cao N.**, **Wang C.**, **Li M.**, **Ren K.**, and **Lou W.**, (2013), Privacy-preserving multi-keyword ranked search over encrypted cloud data, *IEEE Transactions on parallel and distributed systems*, **25**(1), 222–233.

19. **Chen K.**, **Hu C.**, **Zhang X.**, **Zheng K.**, **Chen Y.**, and **Vasilakos A. V.**, (2011), Survey on routing in data centers: insights and future directions, *IEEE network*, **25**(4), 6–10.

20. **Chen M.-H.**, **Tung Y.-C.**, **Hung S.-H.**, **Lin K. C.-J.**, and **Chou C.-F.**, (2015), Availability Is Not Enough: Minimizing Joint Response Time in Peer-Assisted Cloud Storage Systems, *IEEE Systems Journal*, **10**(4), 1424–1434.

21. **Cui L.**, **Zhang J.**, **Yue L.**, **Shi Y.**, **Li H.**, and **Yuan D.**, (2015), A genetic algorithm based data replica placement strategy for scientific applications in clouds, *IEEE Transactions on Services Computing*, **11**(4), 727–739.

22. **Cunsolo V. D.**, **Distefano S.**, **Puliafito A.**, and **Scarpa M.**, Volunteer computing and desktop cloud: The cloud@ home paradigm, *In Eighth IEEE international symposium on network computing and applications*. IEEE, 2009, 134–139.

23. **Cusumano M. A.**, (2010), Cloud computing and SaaS as new computing platforms., *Commun. ACM*, **53**(4), 27–29.

24. **Descher M., Masser P., Feilhauer T., Tjoa A. M.,** and **Huemer D.,** Retaining data control to the client in infrastructure clouds, *In International Conference on Availability, Reliability and Security*. IEEE, 2009, 9–16.

25. **Dillon T., Wu C.,** and **Chang E.,** Cloud computing: issues and challenges, *In 2010 24th IEEE international conference on advanced information networking and applications*. Ieee, 2010*a*, 27–33.

26. **Dillon T., Wu C.,** and **Chang E.,** Cloud computing: issues and challenges, *In 24th IEEE international conference on advanced information networking and applications*. Ieee, 2010*b*, 27–33.

27. **Doelitzscher F., Sulistio A., Reich C., Kuijs H.,** and **Wolf D.,** (2011), Private cloud for collaboration and e-Learning services: from IaaS to SaaS, *Computing*, **91**(1), 23–42.

28. **Durao F., Carvalho J. F. S., Fonseka A.,** and **Garcia V. C.,** (2014), A systematic review on cloud computing, *The Journal of Supercomputing*, **68**(3), 1321–1346.

29. **Fehling C., Leymann F., Retter R., Schupeck W.,** and **Arbitter P.,** *Cloud computing patterns: fundamentals to design, build, and manage cloud applications*. Springer, 2014.

30. **Feng B., Ma X., Guo C., Shi H., Fu Z.,** and **Qiu T.,** (2016), An efficient protocol with bidirectional verification for storage security in cloud computing, *IEEE Access*, **4**, 7899–7911.

31. **Furht B.,** Cloud computing fundamentals, *In Handbook of cloud computing*, Springer, 2010, 3–19.

32. **Gajek S., Liao L.,** and **Schwenk J.,** Breaking and fixing the inline approach, *In Proceedings of the 2007 ACM workshop on Secure web services*. ACM, 2007, 37–43.

33. **Garber L.,** (2012), The challenges of securing the virtualized environment, *Computer*, **45**(1), 17–20.

34. **Gentry C.** and **Boneh D.,** *A fully homomorphic encryption scheme*, volume 20. Stanford University Stanford, 2009.

35. **Gopinath S.** and **Sherly E.,** (2018), A Comprehensive Survey on Data Replication Techniques in Cloud Storage Systems, *International Journal of Applied Engineering Research*, **13**(22), 15926–15932.

36. **Gul I.**, **ur Rehman A.**, and **Islam M. H.**, Cloud computing security auditing, *In The 2nd International Conference on Next Generation Information Technology*. IEEE, 2011, 143–148.

37. **Guo X.**, **Srivastav M.**, **Huang S.**, **Ganta D.**, **Henry M. B.**, **Nazhandali L.**, and **Schaumont P.**, Silicon implementation of sha-3 finalists: BLAKE, Grøstl, JH, Keccak and Skein, *In ECRYPT II Hash Workshop*. 2011.

38. **Han F.**, **Qin J.**, **Zhao H.**, and **Hu J.**, (2014), A general transformation from KP-ABE to searchable encryption, *Future Generation Computer Systems*, **30**, 107–115.

39. **Haoyu L.**, **Qingpeng L.**, **Longjun Z.**, and **Zhiqiang G.**, Integrity Verification Scheme for Dynamic Cloud Data Based on Multi-Branching Tree, *In 9th International Conference on Measuring Technology and Mechatronics Automation (ICMTMA)*. IEEE, 2017, 365–368.

40. **Harauz J.**, **Kaufman L. M.**, and **Potter B.** (2009), Data security in the world of cloud computing published by the IEEE computer and reliability societies in July.

41. **Huang K.** and **Tso R.**, A commutative encryption scheme based on ElGamal encryption, *In International Conference on Information Security and Intelligent Control*. IEEE, 2012, 156–159.

42. **Jensen M.**, **Schwenk J.**, **Gruschka N.**, and **Iacono L. L.**, On technical security issues in cloud computing, *In IEEE International Conference on Cloud Computing*. Ieee, 2009, 109–116.

43. **Jing X.** and **Jian-Jun Z.**, A brief survey on the security model of cloud computing, *In Ninth international symposium on distributed computing and applications to business, engineering and science*. IEEE, 2010, 475–478.

44. **JoSEP A. D.**, **KAtz R.**, **KonWinSKi A.**, **Gunho L.**, **PAttERSon D.**, and **RABKin A.**, (2010), A view of cloud computing, *Communications of the ACM*, **53**(4).

45. **Kalloniatis C.**, **Mouratidis H.**, **Vassilis M.**, **Islam S.**, **Gritzalis S.**, and **Kavakli E.**, (2014), Towards the design of secure and privacy-oriented information systems in the cloud: Identifying the major concepts, *Computer Standards & Interfaces*, **36**(4), 759–775.

46. **Kandukuri B. R.**, **Rakshit A.**, *et al.*,, Cloud security issues, *In IEEE International Conference on Services Computing*. IEEE, 2009, 517–520.

47. **Kaur A.** and **Bhardwaj M.**, (2012), Hybrid encryption for cloud database security, *Journal of Engineering Science Technology*, **2**, 737–741.

48. **Kavis M. J.**, *Architecting the cloud: design decisions for cloud computing service models (SaaS, PaaS, and IaaS).* John Wiley & Sons, 2014.

49. **Khan A. N.**, **Kiah M. M.**, **Madani S. A.**, **Ali M.**, **Shamshirband S.**, *et al.,*, (2014*a*), Incremental proxy re-encryption scheme for mobile cloud computing environment, *The Journal of Supercomputing*, **68**(2), 624–651.

50. **Khan A. N.**, **Kiah M. M.**, **Madani S. A.**, **Ali M.**, **Shamshirband S.**, *et al.,*, (2014*b*), Incremental proxy re-encryption scheme for mobile cloud computing environment, *The Journal of Supercomputing*, **68**(2), 624–651.

51. **Kim A.**, **McDermott J.**, and **Kang M.**, Security and architectural issues for national security cloud computing, *In IEEE 30th International Conference on Distributed Computing Systems Workshops*. IEEE, 2010, 21–25.

52. **Kretzschmar M.**, **Golling M.**, and **Hanigk S.**, Security management areas in the inter-cloud, *In IEEE 4th International Conference on Cloud Computing*. IEEE, 2011, 762–763.

53. **Krutz R. L.** and **Vines R. D.**, *Cloud security: A comprehensive guide to secure cloud computing*. Wiley Publishing, 2010.

54. **Le A.**, **Markopoulou A.**, **Dimakis A. G.**, **Le A.**, **Markopoulou A.**, and **Dimakis A. G.**, (2016), Auditing for distributed storage systems, *IEEE/ACM Transactions on Networking (TON)*, **24**(4), 2182–2195.

55. **Li J.**, (2017), Novel availability and integrity verification protocol for ISMAC system under cloud environment, *Computers & Electrical Engineering*, **57**, 209–219.

56. **Li W.**, **Yang Y.**, and **Yuan D.**, (2015), Ensuring cloud data reliability with minimum replication by proactive replica checking, *IEEE Transactions on Computers*, **65**(5), 1494–1506.

57. **Li Y.**, **Yu Y.**, **Min G.**, **Susilo W.**, **Ni J.**, and **Choo K.-K. R.**, (2017), Fuzzy identity-based data integrity auditing for reliable cloud storage systems, *IEEE Transactions on Dependable and Secure Computing*.

58. **Lin J.-W.**, **Chen C.-H.**, and **Chang J. M.**, (2013), QoS-aware data replication for data-intensive applications in cloud computing systems, *IEEE Transactions on Cloud Computing*, **1**(1), 101–115.

59. **Liu C.**, **Ranjan R.**, **Yang C.**, **Zhang X.**, **Wang L.**, and **Chen J.**, (2014*a*), MuR-DPA: Top-down levelled multi-replica merkle hash tree based secure public auditing for dynamic big data storage on cloud, *IEEE Transactions on Computers*, **64**(9), 2609–2622.

60. **Liu F.**, **Tong J.**, **Mao J.**, **Bohn R.**, **Messina J.**, **Badger L.**, and **Leaf D.**, (2011), NIST cloud computing reference architecture, *NIST special publication*, **500**(2011), 1–28.

61. **Liu Q.**, **Wang G.**, and **Wu J.**, (2014*b*), Consistency as a service: Auditing cloud consistency, *IEEE Transactions on Network and Service Management*, **11**(1), 25–35.

62. **Manivannan D.** and **Sujarani R.**, Light weight and secure database encryption using tsfs algorithm, *In Second International conference on Computing, Communication and Networking Technologies*. IEEE, 2010, 1–7.

63. **Marinos A.** and **Briscoe G.**, Community cloud computing, *In IEEE International Conference on Cloud Computing*. Springer, 2009, 472–484.

64. **Mei L.**, **Chan W. K.**, and **Tse T.**, A tale of clouds: Paradigm comparisons and some thoughts on research issues, *In IEEE Asia-Pacific Services Computing Conference*. Ieee, 2008, 464–469.

65. **Meng S.** and **Liu L.**, (2012), Enhanced monitoring-as-a-service for effective cloud management, *IEEE Transactions on Computers*, **62**(9), 1705–1720.

66. **Mitali V. K.** and **Sharma A.**, (2014), A survey on various cryptography techniques, *International Journal of Emerging Trends & Technology in Computer Science (IJETTCS)*, **3**(4), 307–312.

67. **Mohamed E. M.**, **Abdelkader H. S.**, and **El-Etriby S.**, Enhanced data security model for cloud computing, *In 8th International Conference on Informatics and Systems (INFOS)*. IEEE, 2012, CC–12.

68. **Monfared A. T.** and **Jaatun M. G.**, Monitoring intrusions and security breaches in highly distributed cloud environments, *In IEEE Third International Conference on Cloud Computing Technology and Science*. IEEE, 2011, 772–777.

69. **Moreno-Vozmediano R.**, **Montero R. S.**, and **Llorente I. M.**, (2012), Key challenges in cloud computing: Enabling the future internet of services, *IEEE Internet Computing*, **17**(4), 18–25.

70. **Nahir A.**, **Orda A.**, and **Raz D.**, (2015), Replication-based load balancing, *IEEE Transactions on Parallel and Distributed Systems*, **27**(2), 494–507.

71. **Pearson S.** and **Benameur A.**, Privacy, security and trust issues arising from cloud computing, *In IEEE Second International Conference on Cloud Computing Technology and Science*. IEEE, 2010, 693–702.

72. **Peng S.**, **Zhou F.**, **Wang Q.**, **Xu Z.**, and **Xu J.**, (2017), Identity-based public multi-replica provable data possession, *IEEE Access*, **5**, 26990–27001.

73. **Rahimi M. R.**, **Venkatasubramanian N.**, and **Vasilakos A. V.**, MuSIC: Mobility-aware optimal service allocation in mobile cloud computing, *In 2013 IEEE Sixth International Conference on Cloud Computing*. IEEE, 2013, 75–82.

74. **Rahmani H.**, **Sundararajan E.**, **Ali Z. M.**, and **Zin A. M.**, (2013), Encryption as a Service (EaaS) as a Solution for Cryptography in Cloud, *Procedia Technology*, **11**, 1202–1210.

75. **Ram C. P.** and **Sreenivaasan G.**, Security as a service (sass): Securing user data by coprocessor and distributing the data, *In Trendz in Information Sciences & Computing (TISC2010)*. IEEE, 2010, 152–155.

76. **Rao A.**, (2012), Centralized database security in cloud, *International Journal of Advanced Research in Computer and Communication Engineering*, **1**, 544–549.

77. **Rao R. V.** and **Selvamani K.**, (2015), Data security challenges and its solutions in cloud computing, *Procedia Computer Science*, **48**, 204–209.

78. **Reddy B. A.** and **Reddy P. R. S.**, (2012), Effective data distribution techniques for multi-cloud storage in cloud computing, *CSE, Anurag Group of Institutions, Hyderabad, AP, India*, 1130–1134.

79. **Reddy M.**, **Manjula S.**, and **Venugopal K.**, (2017), Secure data sharing in cloud computing: a comprehensive review, *International Journal of Computer (IJC)*, **25**(1), 80–115.

80. **Ren K.**, **Wang C.**, and **Wang Q.**, (2012), Security challenges for the public cloud, *IEEE Internet Computing*, **16**(1), 69–73.

81. **Rivest R. L., Adleman L., Dertouzos M. L.**, *et al.,*, (1978), On data banks and privacy homomorphisms, *Foundations of secure computation*, **4**(11), 169–180.

82. **Roman H. N., Kounev S.**, and **Reussner R.**, Elasticity in cloud computing: What it is and what it is not, *In Proceedings of the 10th International Conference on Autonomic Computing (ICAC 2013)*. 2013.

83. **Rong C., Nguyen S. T.**, and **Jaatun M. G.**, (2013*a*), Beyond lightning: A survey on security challenges in cloud computing, *Computers & Electrical Engineering*, **39**(1), 47–54.

84. **Rong C., Nguyen S. T.**, and **Jaatun M. G.**, (2013*b*), Beyond lightning: A survey on security challenges in cloud computing, *Computers & Electrical Engineering*, **39**(1), 47–54.

85. **Rong H., Wang H., Liu J., Wu W.**, and **Xian M.**, Efficient Integrity Verification of Secure Outsourced kNN Computation in Cloud Environments, *In IEEE Trustcom/BigDataSE/ISPA*. IEEE, 2016, 236–243.

86. **Ryan M. D.**, (2013), Cloud computing security: The scientific challenge, and a survey of solutions, *Journal of Systems and Software*, **86**(9), 2263–2268.

87. **Seccombe A., Hutton A., Meisel A., Windel A., Mohammed A., Licciardi A.**, *et al.,*, (2009), Security guidance for critical areas of focus in cloud computing, *Cloud Security Alliance*, **2**(1), 2–70.

88. **Sen J.**, Security and privacy issues in cloud computing, *In Cloud Technology: Concepts, Methodologies, Tools, and Applications*, IGI Global, 2015, 1585–1630.

89. **Shamshirband S., Anuar N. B., Kiah M. L. M.**, and **Patel A.**, (2013), An appraisal and design of a multi-agent system based cooperative wireless intrusion detection computational intelligence technique, *Engineering Applications of Artificial Intelligence*, **26**(9), 2105–2127.

90. **Subashini S.** and **Kavitha V.**, (2011), A survey on security issues in service delivery models of cloud computing, *Journal of network and computer applications*, **34**(1), 1–11.

91. **Sun D., Chang G., Sun L.**, and **Wang X.**, (2011), Surveying and analyzing security, privacy and trust issues in cloud computing environments, *Procedia Engineering*, **15**, 2852–2856.

92. **Tabet K., Mokadem R., Laouar M. R.**, and **Eom S.**, (2017), Data replication in cloud systems: a survey, *International Journal of Information Systems and Social Change (IJISSC)*, **8**(3), 17–33.

93. **Takabi H.**, **Joshi J. B.**, and **Ahn G.-J.**, Securecloud: Towards a comprehensive security framework for cloud computing environments, *In IEEE 34th Annual Computer Software and Applications Conference Workshops*. IEEE, 2010, 393–398.

94. **Tebaa M.** and **Hajji S. E.**, (2014), Secure cloud computing through homomorphic encryption, *arXiv preprint arXiv:1409.0829*.

95. **Thosar S. D.** and **Sonewar P.**, Data integrity verification privacy preserving approach of cloud using Third Party Auditor and multithreading, *In 2016 International Conference on Computing Communication Control and automation (ICCUBEA)*. IEEE, 2016, 1–4.

96. **Viega J.**, (2009), Cloud computing and the common man, *Computer*, **42**(8), 106–108.

97. **Wang C.**, **Ren K.**, **Lou W.**, and **Li J.**, (2010), Toward publicly auditable secure cloud data storage services, *IEEE network*, **24**(4), 19–24.

98. **Wang C.**, **Wang Q.**, **Ren K.**, **Cao N.**, and **Lou W.**, (2011), Toward secure and dependable storage services in cloud computing, *IEEE transactions on Services Computing*, **5**(2), 220–232.

99. **Wang H.**, (2012), Proxy provable data possession in public clouds, *IEEE Transactions on Services Computing*, **6**(4), 551–559.

100. **Wang H.**, **He D.**, and **Tang S.**, (2016*a*), Identity-based proxy-oriented data uploading and remote data integrity checking in public cloud, *IEEE Transactions on Information Forensics and Security*, **11**(6), 1165–1176.

101. **Wang L.**, **Tao J.**, **Kunze M.**, **Castellanos A. C.**, **Kramer D.**, and **Karl W.**, Scientific cloud computing: Early definition and experience, *In 10th ieee international conference on high performance computing and communications*. Ieee, 2008, 825–830.

102. **Wang L.**, **Zhang F.**, **Aroca J. A.**, **Vasilakos A. V.**, **Zheng K.**, **Hou C.**, **Li D.**, and **Liu Z.**, (2013), GreenDCN: A general framework for achieving energy efficiency in data center networks, *IEEE Journal on Selected Areas in Communications*, **32**(1), 4–15.

103. **Wang L.**, **Zhang F.**, **Zheng K.**, **Vasilakos A. V.**, **Ren S.**, and **Liu Z.**, Energy-efficient flow scheduling and routing with hard deadlines in data center networks, *In IEEE 34th International Conference on Distributed Computing Systems*. IEEE, 2014, 248–257.

104. **Wang Y.**, **Wu Q.**, **Qin B.**, **Shi W.**, **Deng R. H.**, and **Hu J.**, (2016*b*), Identity-based data outsourcing with comprehensive auditing in clouds, *IEEE transactions on information forensics and security*, **12**(4), 940–952.

105. **Wei G.**, **Vasilakos A. V.**, **Zheng Y.**, and **Xiong N.**, (2010*a*), A game-theoretic method of fair resource allocation for cloud computing services, *The journal of supercomputing*, **54**(2), 252–269.

106. **Wei Q.**, **Veeravalli B.**, **Gong B.**, **Zeng L.**, and **Feng D.**, CDRM: A cost-effective dynamic replication management scheme for cloud storage cluster, *In 2010 IEEE international conference on cluster computing*. IEEE, 2010*b*, 188–196.

107. **Wu L.**, **Garg S. K.**, and **Buyya R.**, SLA-based resource allocation for software as a service provider (SaaS) in cloud computing environments, *In Proceedings of the 2011 11th IEEE/ACM International Symposium on Cluster, Cloud and Grid Computing*. IEEE Computer Society, 2011, 195–204.

108. **Xu X.**, **Yang C.**, and **Shao J.**, (2017), Data replica placement mechanism for open heterogeneous storage systems, *Procedia Computer Science*, **109**, 18–25.

109. **Yang K.** and **Jia X.**, (2012), An efficient and secure dynamic auditing protocol for data storage in cloud computing, *IEEE transactions on parallel and distributed systems*, **24**(9), 1717–1726.

110. **Yu Y.**, **Au M. H.**, **Ateniese G.**, **Huang X.**, **Susilo W.**, **Dai Y.**, and **Min G.**, (2016*a*), Identity-based remote data integrity checking with perfect data privacy preserving for cloud storage, *IEEE Transactions on Information Forensics and Security*, **12**(4), 767–778.

111. **Yu Y.**, **Au M. H.**, **Ateniese G.**, **Huang X.**, **Susilo W.**, **Dai Y.**, and **Min G.**, (2016*b*), Identity-based remote data integrity checking with perfect data privacy preserving for cloud storage, *IEEE Transactions on Information Forensics and Security*, **12**(4), 767–778.

112. **Zhang J.**, **Liu S.**, and **Mao J.**, On the security of a dynamic-hash-table based public auditing protocol, *In 9th International Congress on Image and Signal Processing, BioMedical Engineering and Informatics (CISP-BMEI)*. IEEE, 2016*a*, 1954–1958.

113. **Zhang L.-J.** and **Zhou Q.**, CCOA: Cloud computing open architecture, *In IEEE International Conference on Web Services*. Ieee, 2009, 607–616.

114. **Zhang N.**, **Liu D.**, and **Zhang Y.**, A research on cloud computing security, *In International Conference on Information Technology and Applications*. IEEE, 2013, 370–373.

115. **Zhang Q.**, **Cheng L.**, and **Boutaba R.**, (2010), Cloud computing: state-of-the-art and research challenges, *Journal of internet services and applications*, **1**(1), 7–18.

116. **Zhang X.**, **Zhou F.**, **Zhu X.**, **Sun H.**, **Perrig A.**, **Vasilakos A. V.**, and **Guan H.**, (2014), DFL: Secure and practical fault localization for datacenter networks, *IEEE/ACM Transactions on Networking (TON)*, **22**(4), 1218–1231.

117. **Zhang Y.**, **Xu C.**, **Li H.**, and **Liang X.**, (2016*b*), Cryptographic public verification of data integrity for cloud storage systems, *IEEE Cloud Computing*, **3**(5), 44–52.

118. **Zhang Y.**, **Xu C.**, **Liang X.**, **Li H.**, **Mu Y.**, and **Zhang X.**, (2016*c*), Efficient public verification of data integrity for cloud storage systems from indistinguishability obfuscation, *IEEE Transactions on Information Forensics and Security*, **12**(3), 676–688.

119. **Zhao W.**, **Liu J.**, and **Hara T.**, (2018), Optimal replica distribution in edge-node-assisted Cloud-P2P Platforms for real-time streaming, *IEEE Transactions on Vehicular Technology*, **67**(9), 8637–8646.

120. **Zhou M.**, **Mu Y.**, **Susilo W.**, **Yan J.**, and **Dong L.**, (2012), Privacy enhanced data outsourcing in the cloud, *Journal of Network and Computer Applications*, **35**(4), 1367–1373.

121. **Zhou M.**, **Zhang R.**, **Xie W.**, **Qian W.**, and **Zhou A.**, Security and privacy in cloud computing: A survey, *In Sixth International Conference on Semantics, Knowledge and Grids*. IEEE, 2010, 105–112.

122. **Zissis D.** and **Lekkas D.**, (2012), Addressing cloud computing security issues, *Future Generation computer systems*, **28**(3), 583–592.

www.ingramcontent.com/pod-product-compliance
Lightning Source LLC
LaVergne TN
LVHW082348200926
848481LV00012B/3498